Water Resources Planning for the Upper Mississippi River and Illinois Waterway

I0807450

Committee to Review the Corps of Engineers Restructured Upper Mississippi River-Illinois Waterway Draft Feasibility Study

Water Science and Technology Board
Division on Earth and Life Studies

Transportation Research Board

NATIONAL RESEARCH COUNCIL
OF THE NATIONAL ACADEMIES

THE NATIONAL ACADEMIES PRESS
Washington, D.C.
www.nap.edu

THE NATIONAL ACADEMIES PRESS 500 Fifth Street, N.W. Washington, DC 20001

NOTICE: The project that is the subject of this report was approved by the Governing Board of the National Research Council, whose members are drawn from the councils of the National Academy of Sciences, the National Academy of Engineering, and the Institute of Medicine. The members of the committee responsible for the report were chosen for their special competences and with regard for appropriate balance.

This study was supported by Contract No. DACW17-03-C-0003 between the National Academy of Sciences and the Department of the Army. Any opinions, findings, conclusions, or recommendations expressed in this publication are those of the author(s) and do not necessarily reflect the views of the organizations or agencies that provided support for the project.

International Standard Book Number 0-309-009945-5

Water Resources Planning for the Upper Mississippi River and Illinois Waterway is available from National Academies Press, 500 Fifth Street, N.W., Lockbox 285, Washington, DC 20055; (800) 624-6242 or (202) 334-3313 (in the Washington metropolitan area); Internet, http://www.nap.edu.

Cover photo credits: bald eagle by Thomas O'Neil and lock and dam courtesy of the Alexis Park Inn & Suites, Iowa City, Iowa. Back cover map taken from *Views on the Mississippi, the Photographs of Henry Peter Bosse*, reproduced with permission, from the John R. Borchert Map Library, University of Minnesota.

Printed in the United States of America

THE NATIONAL ACADEMIES

Advisers to the Nation on Science, Engineering, and Medicine

The **National Academy of Sciences** is a private, nonprofit, self-perpetuating society of distinguished scholars engaged in scientific and engineering research, dedicated to the furtherance of science and technology and to their use for the general welfare. Upon the authority of the charter granted to it by the Congress in 1863, the Academy has a mandate that requires it to advise the federal government on scientific and technical matters. Dr. Ralph J. Cicerone is president of the National Academy of Sciences.

The **National Academy of Engineering** was established in 1964, under the charter of the National Academy of Sciences, as a parallel organization of outstanding engineers. It is autonomous in its administration and in the selection of its members, sharing with the National Academy of Sciences the responsibility for advising the federal government. The National Academy of Engineering also sponsors engineering programs aimed at meeting national needs, encourages education and research, and recognizes the superior achievements of engineers. Dr. Wm. A. Wulf is president of the National Academy of Engineering.

The **Institute of Medicine** was established in 1970 by the National Academy of Sciences to secure the services of eminent members of appropriate professions in the examination of policy matters pertaining to the health of the public. The Institute acts under the responsibility given to the National Academy of Sciences by its congressional charter to be an adviser to the federal government and, upon its own initiative, to identify issues of medical care, research, and education. Dr. Harvey V. Fineberg is president of the Institute of Medicine.

The **National Research Council** was organized by the National Academy of Sciences in 1916 to associate the broad community of science and technology with the Academy's purposes of furthering knowledge and advising the federal government. Functioning in accordance with general policies determined by the Academy, the Council has become the principal operating agency of both the National Academy of Sciences and the National Academy of Engineering in providing services to the government, the public, and the scientific and engineering communities. The Council is administered jointly by both Academies and the Institute of Medicine. Dr. Ralph J. Cicerone and Dr. Wm. A. Wulf are chair and vice chair, respectively, of the National Research Council.

www.national-academies.org

COMMITTEE TO REVIEW THE CORPS OF ENGINEERS RESTRUCTURED UPPER MISSISSIPPI RIVER-ILLINOIS WATERWAY DRAFT EASIBILITY STUDY[1]

JOHN J. BOLAND, *Chair*, Johns Hopkins University (Emeritus), Baltimore, Maryland
PATRICK BREZONIK, University of Minnesota, Minneapolis
ROBERT K. DAVIS, University of Colorado, Boulder (retired)
LEO M. EISEL, Brown and Caldwell, Golden, Colorado
STEPHEN W. FULLER, Texas A&M University, College Station
GERALD E. GALLOWAY, University of Maryland, College Park
LESTER B. LAVE, Carnegie Mellon University, Pittsburgh, Pennsylvania
KARIN LIMBURG, Syracuse University, New York
ELIZABETH A. RIEKE, U.S. Bureau of Reclamation, Carson City, Nevada
SOROOSH SOROOSHIAN, University of California, Irvine
RICHARD E. SPARKS, National Great Rivers Research and Education Center, Alton, Illinois

NRC Staff

JEFFREY W. JACOBS, Study Director, Water Science and Technology Board
JOSEPH MORRIS, Senior Staff Officer, Transportation Research Board
ANITA A. HALL, Program Associate, Water Science and Technology Board

Editor

Jim Lawson

[1]The activities of the Committee to Review the Corps of Engineers Restructured Upper Mississippi River-Illinois Waterway Draft Feasibility Study are overseen and supported by the National Research Council's Water Science and Technology Board (lead) and the Transportation Research Board. WSTB and TRB members are listed in Appendix B. Biographical information on committee members is contained in Appendix C.

Preface

This is the third and final report from the National Research Council (NRC) Committee to Review the Corps of Engineers Restructured Upper Mississippi River-Illinois Waterway Feasibility Study. The committee was formed in mid-2003. Since then, committee members have reviewed Corps documents and supporting studies, attended meetings and briefings, spoke with Corps staff members and other analysts involved in preparation of the feasibility study, and prepared reports. In the course of this work the committee had the opportunity to learn many things about the commerce, ecology, and history of the Upper Mississippi River-Illinois Waterway (UMR-IWW) region. That learning process included Mississippi River outings on towboats and flat-bottom aluminum boats, and visits to sandbars, backwaters, and floodplains along the river.

The NRC's review of the UMR-IWW Feasibility Study has been a lengthy engagement by any measure. It involved two committees and required more than five years to bring the review to this point. The first (Phase I) committee was convened in early 2000 to conduct a limited review of a feasibility study that was thought to be nearing completion. In its report, issued in 2001, the Phase I committee criticized several important aspects of the Corps study. This report, as well as criticisms received from other sources, led the Corps to pause the planning effort and to restructure the study before proceeding. Once work was well underway on the Restructured UMR-IWW Feasibility Study, the Corps invited the NRC to form a second (Phase II) review committee, asking it to perform an independent technical review of the entire feasibility study. The Phase II committee was formed in 2003 and met for the first time in September of that year in Washington, D.C. That meeting was devoted to reviewing of the first product of the restructured planning effort, an interim feasibility report. This review is documented in the committee's first report (NRC, 2004a).

Subsequent meetings in St. Louis, Missouri; Irvine, California; and Red Wing, Minnesota, reviewed the evolving final feasibility study. The committee presented its conclusions and recommendations in a second report (NRC, 2004b). At that point the Phase II NRC review moved on to the subject matter for the third and last report. After a final meeting in Washington, D.C., the committee prepared the following report. The statement of task for the committee's first two reports were identical, calling for the committee to review the feasibility study's economic, engineering, and environmental aspects. This report has a different purpose. It considers larger planning and management issues in the UMR-IWW and within the Corps of Engineers, and also considers UMR-IWW management in the context of several NRC reports on Corps of Engineers' planning procedures. Although the committee was at times tempted to revisit and comment upon its first two reports, this third report does not focus on ground that was covered in the committee's first two reports. It is also worth mentioning that the statement of task for this third report differs from the statement of task that was earlier envisioned. As stated in this committee's first report, the committee's third report was originally intended to "review the Corps' responses to advice from the NRC Upper Mississippi River-Illinois Waterway studies (i.e., the reports of this committee and the NRC 2001 Phase I committee)" (NRC, 2004a). As this committee approached its third and final report, the Corps and the NRC agreed that a more useful and appropriate charge for this committee's third report would focus on the NRC's recent review of the Corps of Engineers' planning methods and approaches. These NRC reports, also known as the "216" studies, reviewed several planning concepts relevant to UMR-IWW resources management and were used as a point of departure for this committee's third report.

Throughout this extended review the committee has benefited greatly from a working relationship with Corps staff that has been as helpful and cooperative as it has been professional and mindful of the special responsibilities of independent reviewers. We particularly appreciate the effort that the Corps made to support this last part of our investigation. Among those participating in the Washington meeting were Major General Don T. Riley, Corps of Engineers Chief of Civil Works; William Dawson, Chief, Corps of Engineers Planning and Policy Division; Ken Barr, Environmental Analysis Team Manager, UMR-IWW Navigation Feasibility Study; Denny Lundburg, Chief, Engineering Division, Corps of Engineers Rock Island District; Rich Manguno, Economics Manager, UMR-IWW Navigation Feasibility Study; Chuck Spitzack, Regional Project Manager, UMR-IWW Navigation and Environmental Sustainability Program; and Richard Worthington, Senior Navigation Policy Advisor and Headquarters Manager, UMR-IWW Navigation Feasibility Study. Marcus Peacock from the Office of Management and Budget, and John Paul Woodley, Assistant Secretary of the Army for Civil Works, also joined the meeting, generously sharing their time and thoughts.

The committee's NRC Study Director Jeffrey W. Jacobs coped with the heavy work load that characterized this study process from the beginning, liaising with Corps staff, planning meetings, assembling documents, and—most of all—sharing with the committee chair the responsibility for assembling, editing, and revising committee reports. Jeff continued to be ably backed up by Joseph R. Morris of the NRC's Transportation Research Board, who offered his own expertise to the committee when needed and assisted with editing reports. The Water Science and Technology Board staff, directed by Stephen Parker, provided their usual seamless logistical support, much of it delivered by Senior Project Assistant Anita A. Hall.

This report was reviewed in draft form by individuals chosen for their diverse perspectives and technical expertise in accordance with the procedures approved by the NRC's Report Review Committee. The purpose of this independent review is to provide candid and critical comments that will assist the institution in making its published report as sound as possible and to ensure that the report meets institutional standards for objectivity, evidence, and responsiveness to the study charge. The review comments and draft manuscript remain confidential to protect the integrity of the deliberative process. We thank the following individuals for their review of this report:

Donald F. Boesch, University of Maryland Center for Estuarine Studies
Douglas M. Johnston, University of Illinois
Catherine L. Kling, Iowa State University
Kai N. Lee, Williams College
Daniel P. Loucks, Cornell University
Nicholas Pinter, Southern Illinois University
Leonard Shabman, Resources for the Future
M. Gordon Wolman, Johns Hopkins University

Although the reviewers listed above have provided many constructive comments and suggestions, they were not asked to endorse the conclusions or recommendations, nor did they see the final draft of the report before its release. The review of this report was overseen by Walter R. Lynn, Cornell University, who was appointed by the Division on Earth and Life Studies and by Frank H. Stillinger, Princeton University, who was appointed by the NRC's Report Review Committee. Drs. Lynn and Stillinger were responsible for making certain that an independent examination of this report was carried out in accordance with institutional procedures and that all review comments were carefully considered. Responsibility for the final content of this report rests entirely with the authoring committee and the institution.

As the nation's river engineering agency, the Corps of Engineers will continue to play a lead role in operating and maintaining the UMR-IWW navigation project and its related benefits and uses. As societal values and resource availabilities continue to shift in unknown ways in the future, the configuration and operations of the UMR-IWW navigation system may likewise change. But it seems clear that the Corps will continue the trend toward working more closely with other federal agencies, state resources agencies, and citizens along the river. The Corps made conspicuous gains in this area during the conduct of the UMR-IWW study. This collaboration is a welcome development, as were a number of other innovations and achievements described in this report.

Our committee was pleased to play a role in this process. Just as it is hoped that the committee's reports will have some lasting value in future river management decisions, we are each individually wiser for having been part of this process. We learned a great deal from meeting and working with the many Corps of Engineers employees, other federal employees, state employees, commercial navigators, nongovernmental organizations, and numerous citizens along the river. We wish them well in their future collaborations and as they work toward enjoying and managing the immense national resources of the Upper Mississippi River and the Illinois Waterway.

John J. Boland, *Chair*

Contents

Summary

In the late 1980s the U.S. Army Corps of Engineers began considering the possibility of extending several locks on the Upper Mississippi River-Illinois Waterway (UMR-IWW) as a means for reducing waterway traffic congestion. A reconnaissance study, begun in 1989, led to a feasibility study, which began in 1993 and continued for the next 11 years. As the UMR-IWW feasibility study progressed, the Corps confronted several particularly difficult technical, organizational, and political challenges. The planning process finally ended in December 2004 when the Chief of Engineers, Lieutenant General Carl Strock, sent the final Chief's Report to the Assistant Secretary of the Army for Civil Works for review prior to submission to Congress. The recommended plan included a $5.3 billion program for ecosystem restoration and a $2.4 billion program for navigation infrastructure improvements to be expended over the next several decades.

During the course of the feasibility study, the Corps developed and applied several novel planning approaches and methods. There was an effort to develop:

- a spatial equilibrium model for grain shipments and prices (the ESSENCE model);
- a federal interagency Principals Group was established in 2001 to enhance communication among the Corps and other federal agencies with UMR-IWW management responsibilities;
- a regional coordinating group was established, involving federal agencies, state resources agencies, and nongovernmental organizations;
- the Corps commissioned a report that presented several scenarios of future U.S. grain exports; and
- the Corps explored means for adaptive management on the UMR-IWW, not only for ecosystem restoration activities but also for determining the timing and extent of navigation-related construction activities.

To obtain independent technical review of the evolving feasibility study, the Corps sought the advice of two National Research Council (NRC) committees. A first, Phase I, committee was established in 2000 to review the economic aspects of the draft feasibility study. That committee issued a single report in early 2001. A second, Phase II, committee was convened in 2003 to provide a more comprehensive technical review of the ongoing feasibility study, which had been broadened to include a significant ecosystem restoration component. The Phase II committee issued two reports in 2004: a first report that provided the committee's initial impressions of the draft feasibility study and a second report that constituted a more in-depth review.

This is the third and final report from the NRC Phase II committee. This committee's first two reports focused on the analytical and technical aspects of the Corps feasibility study. This third report assumes a different point of view, as it considers several broader issues that affected the Corps' feasibility study. The statement of task for this third report also differs from the statement of task that was envisioned at the outset of this project. As stated in this committee's first report, the committee's third report was originally intended to "review the Corps' responses to advice from the NRC Upper Mississippi River-Illinois Waterway studies (i.e., the reports of this committee and the NRC 2001 Phase I committee)" (NRC, 2004a). As this committee approached its third and final report, the Corps and the NRC agreed that a more useful and appropriate charge for this committee's third report would focus on the NRC's recent review of the Corps of Engineers' planning methods and approaches.

These NRC studies are referred to as the "216" studies, as they were conducted in response to a congressional request in Section 216 of the Water Resources Development Act of 2000 (WRDA 2000). Five reports on Corps of Engineers planning were issued as part of the 216 study activity: peer review (2002), adaptive management, analytical methods, river basins and coastal systems, and a report from a "coordinating committee." These latter four reports were all issued in 2004; Chapter 2 of this committee's report reviews and summarizes all five 216 study reports.

The committee was requested to consider the 216 studies, as well as other planning issues that affect UMR-IWW system management. Accordingly, this report addresses the following topics: interagency coordination, UMR-IWW authorities and legislation, adaptive management, valuation of nonmarket benefits and costs, and streamlining of Corps planning studies. The report also discusses the importance of coordination between the Institute of Water Resources (IWR) and district planners, and the roles of peer review in the feasibility study process; however, no formal recommendations are offered on these subjects. The formal statement of task for this report is included verbatim in Chapter 1.

INTERAGENCY COORDINATION

To promote interagency coordination among the key federal agencies with UMR-IWW responsibilities, a Principals Group consisting of senior-level representation from the Corps, the Department of Agriculture, the Department of Transportation, the Environmental Protection Agency, and the Fish and Wildlife Service was established in 2001. This group was formed to ensure that changes and updates in the feasibility study were being shared in an interagency forum. Given the importance of the Principals Group, and given that other, similar bodies have been established to promote interagency collaboration in other U.S. river and aquatic systems, a formal, independent review of how the Principals Group affected the feasibility study process would be useful. Part of this review should consider similarities and differences between the UMR-IWW Principals Group and other interagency water system management groups across the nation (examples include California's CALFED program, the Glen Canyon Adaptive Management Program, the Louisiana Coastal Area Ecosystem Restoration Project, and the South Florida Ecosystem Restoration Task Force). The experiences from the Principals Group should be useful to future interagency cooperation on the UMR-IWW, and should be of interest to other such forums in the United States and in other regions of the world.

The Corps of Engineers should enlist the services of an independent investigator or a small group of investigators to review and assess the experience with the federal, interagency Principals Group for the UMR-IWW feasibility study. The investigator report should also consider experiences with other high-level interagency groups that have been assembled to help manage large U.S. river and aquatic systems.

UMR-IWW AUTHORITIES AND LEGISLATION

A key observation within the 216 study panel reports was that the Corps of Engineers today makes its planning decisions in accord with a large and diverse body of policies, legislation, executive branch guidance, regulations, and case law that constitute a *de facto* national water policy. In regard to the UMR-IWW, prominent examples of this large body of legislation include the 1930 Rivers and Harbors Act that authorized the 9-foot channel project on the Upper Mississippi River, the Endangered Species Act (ESA) of 1973, the Upper Mississippi River Management Act of 1986, and the numerous Water Resources Development and Flood Control Acts that authorized various water- and flood-related projects.

As pointed out in this committee's second report (NRC, 2004b), these multiple directives for Upper Mississippi and Illinois River operations are not always fully consistent with one another. This often requires the Corps to choose which authorization(s)—and therefore which group of river resource users—are to receive priority. These multiple acts and authorizations also have different implications for channel depths and river flows. An authorization to maintain a minimum 9-foot channel is clear enough; but other authorized purposes, such as the protection of endangered species habitat or improvements in river system ecology, may entail different river channel depths or river flows. Within this muddled legislative setting the Corps generally interprets the 1930 authorization for the 9-foot channel as the overriding authority in managing the UMR-IWW system.

Although some may argue that this type of policy ambiguity provides flexibility for the administration and the Congress in dealing with the multiple constituencies with interests in the UMR-IWW, the existence of conflicting directives places the Corps of Engineers—an executive branch line agency—in the uncomfortable position of choosing which constituency is to receive priority. Moreover, the primacy the Corps accords to the 9-foot channel in UMR-IWW management decisions effectively rules out several potential trade-offs (e.g., maintaining a navigation channel less than nine feet deep at some times of the year) between the commercial navigation sector and other related uses, such as boating and commercial and recreational fishing.

To help the Corps of Engineers and other federal and state agencies better manage and understand the federal intent for use of UMR-IWW resources, the administration and the Congress should clarify relative priorities among the multiple laws, executive branch guidance, and congressional reports that govern UMR-IWW management.

ADAPTIVE MANAGEMENT

The Corps of Engineers made a strong effort to include adaptive management as an element within the UMR-IWW feasibility study. The Corps has employed adaptive processes in its manipulations of pool levels for both navigation system management and ecosystem restoration, and the agency should explore means to continue to learn from experience. There are many opportunities for the Corps to apply adaptive learning principles on the UMR-IWW. Examples are improving waterway traffic forecasting by comparing projections with actual flows, and by applying and learning from nonstructural measures to better manage waterway traffic congestion. Adaptive management is not simply a line item in a project budget or something to be implemented or set aside according to available resources, but rather a resources management approach that should become a central part of the Corps' mode of operations. The attainment of more adaptive UMR-IWW operations and decisions will require resources, and it will require cooperation among the Corps and other federal and state agencies. An adaptive management approach seeks to reduce uncertainties and better understand the complexities of ecosystem management; it also recognizes that mistakes will be made along the way and that adaptive management is not a perfect solution to managing water and related resources. It does, however, hold promise in helping to better understand ecosystem dynamics, connections with social and economic systems, and in identifying more flexible management decisions and regimes. It also encourages retrospective comparisons and evaluations in order to enhance system learning and decision making.

In moving forward with UMR-IWW adaptive management actions, it should be recognized that adaptive management is not a project add-on to be implemented, limited, or set aside according to budgetary constraints. Rather, it is a process and perspective that should become part of the organizational fabric. The administration and the Congress should support the Corps in its efforts to integrate adaptive concepts into the operations of its entire UMR-IWW portfolio, including ecosystem restoration projects, transportation infrastructure, and waterway traffic management. Retrospective comparisons and studies can improve future forecasts and other aspects of UMR-IWW decision making and should be seen as integral to an adaptive approach.

VALUATION OF NONMARKET BENEFITS AND COSTS

Many Corps of Engineers water resources projects deliver benefits and entail costs that cannot be easily measured in monetary terms. When projects have significant benefits or costs that have not been expressed in monetary units, the guidance provided by benefit-cost analysis to justify projects, rank alternatives, or allocate project funds—as required under current project evaluation guidelines—is often distorted. As a result, projects with large monetary benefits (e.g., conventional single-purpose water resource projects) are likely to be favored over those with large nonmonetary benefits (e.g., ecosystem restoration projects), regardless of their ultimate value to society.

In some instances within Corps planning studies the valuation of nonmarket benefits—namely, those from navigation improvements (congestion reduction)—has long been conducted and accepted. On the other hand, methods for the valuation of other types of nonmarket benefits—such as ecosystem restoration—are not typically applied in Corps planning studies. This is despite the fact that theoretically sound, credible methods exist for valuing some benefits of ecosystem restoration in monetary units, which allows for at least partial valuation of resources in most instances. As scientific knowledge accumulates, and in particular as more is learned about linkages between ecosystem functions and services, opportunities for valuation of ecosystem services will expand.

The current state of ecosystem science and economic analysis clearly supports valuation of the benefits of ecosystem restoration. In some cases the valuation of these benefits can be as complete as with the benefits derived in more traditional projects, such as flood control and navigation projects. Even the valuation of some ecosystem restoration benefits will improve the quality of decision making for these projects. In many cases valuation of the most obvious benefits will be sufficient to demonstrate feasibility. In other cases even incomplete valuation may allow for a credible comparison of the remaining nonmonetary benefits against net monetary costs. In either case there is no reason to continue restricting applications of nonmarket methods to traditional categories. All Corps water project benefits and costs should be valued in monetary terms to the extent possible.

STREAMLINING CORPS PLANNING STUDIES

The length of the UMR-IWW feasibility report, including appendices, was measured in the thousands of pages. Documents of this size, written to address all necessary statutory and other guidance, tend to be cumbersome, poorly organized, overly ambitious, and difficult to fully comprehend. Some of these organizational and presentation problems stem from attempts to merge a complex planning study with a complex environmental impact study. Other problems arose from a failure to clearly delineate the scope of the study, which led to efforts to add large amounts of information and data that were of limited relevance. Not only did this make the study difficult to comprehend, but the size and poor organization of the report surely made it difficult for Corps staff to quickly edit and update documents through the course of the study.

The statute-based environmental impact assessment should not be confused with environmental issues and analyses that are integral to a water resources planning study and should be addressed therein. Considering the interplay between economic and environmental (and engineering and social) issues within a feasibility study, however, is a process separate from the conduct of an environmental impact statement. The former is part of a sound water resources planning study, while the latter is conducted pursuant to federal statute to determine the environmental impact of a proposed federal action.

To streamline the preparation of future complex Corps planning reports like the UMR-IWW feasibility study, to enhance their presentation, and to improve their readability, the following steps should be taken:

- There should be a succinct and substantive summary of the key planning issues addressed in the report.
- Although it is essential to integrate economic and environmental issues in to water project plans, the practice of merging of water project planning reports with environmental impact reports should be reconsidered. If these reports are to be merged, the process of integrating them should focus on presenting a clear understanding of the overall report, as well as linkages among its main components. Each of these types of reports presents considerable preparation and presentation complications, however, and in large, complex planning studies, the separation of the environmental impact assessment report from the feasibility and project planning reports should be considered.
- For studies of this magnitude, a full-time staff—technical editor(s) and technical writer(s)—should be retained to oversee the report production and presentation process, including the display of Web-based documents.
- Technical details of planning studies should not be included in the report's main body but should be included in appendices, both on paper and in Web-based documents.

1

Introduction

For well over one hundred years the U.S. Army Corps of Engineers has been responsible for the construction, operation, and maintenance of a major inland waterway system on the Upper Mississippi River (UMR) and the various waterbodies in the State of Illinois that make up the Illinois Waterway (IWW). The UMR-IWW Navigation Project presently uses 35 dams to maintain a navigation channel of at least 9 feet in depth throughout the system; locks are provided at 37 sites, 29 on the UMR and 8 on the IWW (see Figure 1.1). Much of the existing infrastructure was constructed in the 1930s, pursuant to the 1930 Rivers and Harbors Act that authorized a 9-foot channel project for the Upper Mississippi River.

Waterway congestion and the rehabilitation of UMR-IWW locks and dams are issues that date back several decades. For instance, the Corps was studying possibilities for improvements to Upper Mississippi River locks and dams in the 1960s. Waterway traffic congestion at Lock and Dam 26, near Alton, Illinois, led to a recommendation for a new dam with a 1,200-foot lock chamber. This facility, which was renamed Melvin Price Lock and Dam, was completed in 1990. By the late 1980s, however, waterway traffic congestion at other downstream locks and on the Illinois Waterway prompted the Corps to begin evaluating the feasibility of extending other lock chambers from 600 to 1,200 feet in length. This construction was proposed to reduce lockage times by allowing the largest tows (typically consisting of 15 barges plus a towboat) to be processed in a single lockage. The Corps' formal feasibility study of the UMR-IWW began in 1993 and was completed in 2004. The various studies that were conducted as part of this effort were in many ways unprecedented in scope, approaches employed, and in their analytical complexities. As the overall feasibility study progressed, two successive committees of the National Research Council (NRC) were convened to provide independent advice on the feasibility study's various technical aspects.

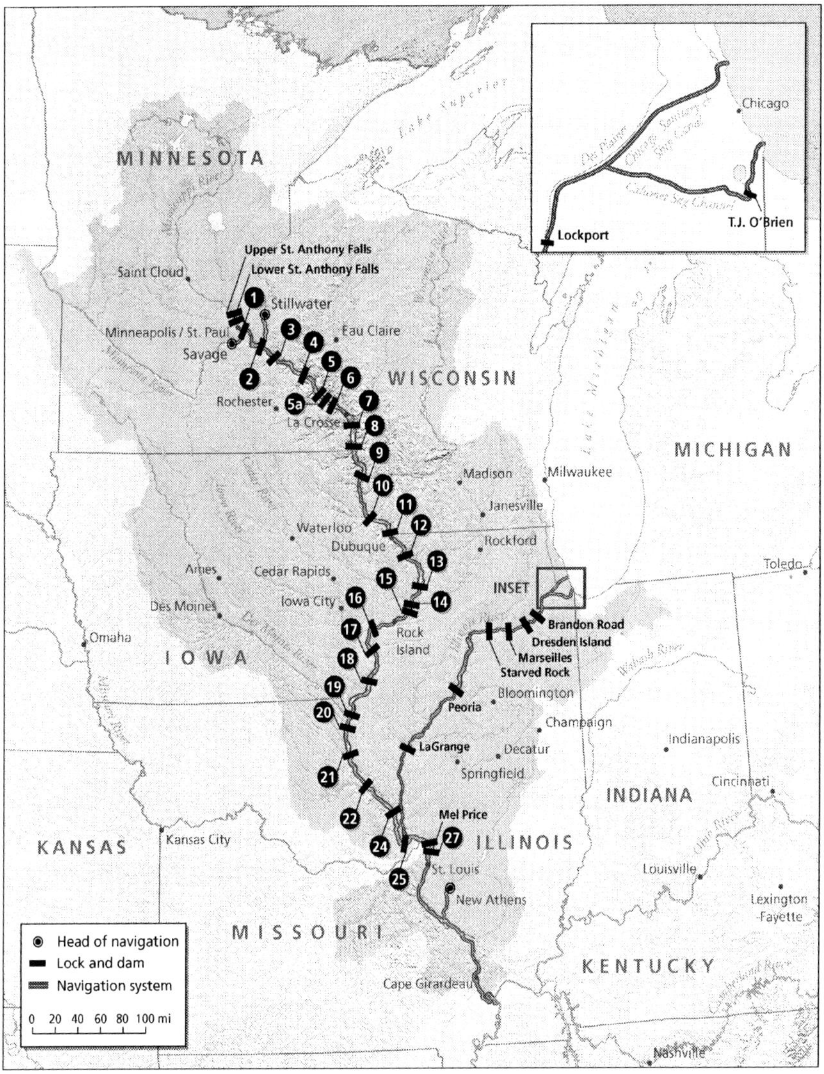

FIGURE 1.1 Locks and dams on the Upper Mississippi River-Illinois Waterway.

The first, or Phase I, committee was convened in 2000 and issued a single report in 2001. At that time the feasibility study was solely concerned with navigation improvements. Subsequent to the issue of the 2001 report, the Corps restructured the feasibility study, leading to a broader and significantly more complex study. Among the major changes was the addition of ecosystem restoration as a second project objective. After resuming work on the restructured feasibility study, the Corps requested the National Research Council to convene a second committee to review technical aspects of the ongoing study. The second, or Phase II, committee began its activities in 2003, issuing two reports in 2004 (NRC, 2004a,b). This is the third and final report from the Phase II committee. This committee's first two reports focused on the ecologic, economic, and engineering aspects of the Corps feasibility study. The charge to the committee was the same for each of these 2004 reports.

This third report is guided by a statement of task that assumes a broader perspective on Corps of Engineers planning and management of UMR-IWW resources (see Box 1.1). Although this report's perspective thus differs from the committee's first two reports, those two reports constitute important background for this third report, and they should be consulted by those seeking deeper understanding of the Corps feasibility study and UMR-IWW management issues such as waterway system management and forecasting, spatial modeling of global and regional grain markets, and ecosystem restoration.

In addition to this committee's two 2004 reports this report builds upon other National Research Council studies. Specifically, the statement of task in Box 1.1 of this report called for the Corps UMR-IWW feasibility study to be considered in the context of the National Research Council's "216" studies. This set of studies, most of which were completed in 2004, were conducted pursuant to Section 216 of the federal Water Resources Development Act (WRDA) of 2000, which requested the NRC to review Corps of Engineers peer review procedures and methods of analysis (Section 216 from WRDA 2000 is reproduced in Appendix A). In response to that congressional directive the NRC convened four separate committees and the Coordinating Committee that followed progress of the individual panels. These five groups issued reports on the topics of:

- Peer review (NRC, 2002);
- Adaptive management (NRC, 2004c);
- Analytical methods (NRC, 2004d);
- River basin and coastal systems (NRC, 2004e); and
- Coordinating Committee (NRC, 2004f).

BOX 1.1
Committee to Review the Corps of Engineers Restructured Upper Mississippi River-Illinois Waterway Feasibility Study

Third Report Statement of Task

The nature of this committee's third and final report will differ from its first two reports. This committee's first two reports focused on reviewing the analyses performed within the Corps Restructured feasibility study for the Upper Mississippi River-Illinois Waterway. This third report will consider several larger issues that affect the conduct of the feasibility study and that affect UMR-IWW system management. The committee may issue additional comments on the Corps feasibility study, but the report's emphasis will be on the following (and perhaps other) larger issues:

- Implications of the 2004 NRC "216 Studies" (that reviewed Corps of Engineers methods of peer review and methods of analysis) for managing resources of the UMR-IWW, and similar large river systems;
- Other water-related issues in the region that relate to integrated UMR-IWW system management, such as water quality and floodplain management;
- Implementing adaptive management; and
- Quantification and valuation of Corps project benefits, especially environmental benefits.

Chapter 2 of this report summarizes the 216 study activity and its five reports. Chapter 3 builds upon the topics presented in Chapter 2 and comments on key 216 study findings and their application to the UMR-IWW feasibility study and system management. Consistent with this committee's statement of task, Chapter 3 discusses the topics of adaptive management and valuation of environmental benefits, along with other topics and reports relevant to UMR-IWW resources management. The report concludes with a short epilogue presented as Chapter 4.

2

National Research Council "216" Studies: Corps of Engineers Water Resources Project Planning

In Section 216 of the Water Resources Development Act (WRDA) of 2000, the U.S. Congress requested The National Academies[2] to review the Corps' methods of project review and analysis (see Appendix A for the exact Section 216 language). In response to this request the Water Science and Technology Board of the National Research Council (NRC), in collaboration with the NRC's Ocean Studies Board, appointed four study committees: (1) peer review; (2) adaptive management; (3) analytical methods; and (4) river basin and coastal systems planning. A coordinating committee was also convened to follow the committees' progress and to issue its own report. This chapter reviews common themes identified in those reports and then summarizes conclusions and recommendations from each of the five reports.

COMMON THEMES WITHIN THE 216 STUDIES

Several common themes regarding Corps of Engineers planning and analytical processes emerged from the five 216 study reports. Given the prominence of the 216 studies within this report's statement of task, key findings and recommendations from those studies are presented in this chapter. This summary is based on the committee's reading of the 216 study reports and is provided as a convenience to the reader who may not be familiar with the reports from the

[1]The National Academies consists of the National Academy of Sciences, the National Academy of Engineering, and the Institute of Medicine. The National Research Council is the research and operating arm of The National Academies.

various 216 study panels, as well as important background information for discussion in Chapter 3. This chapter only summarizes the 216 reports and does not reflect additional critique or commentary from this NRC committee.

The following were key findings from the 216 study panel reports:

- The value of more thorough analyses and peer review during early stages of Corps planning studies. Recommendations in this area included the occasional need for more resources in Corps reconnaissance studies and the potential value of including stakeholder groups in a planning study's initial stages.
- The need for increased postconstruction evaluations, or *ex post* studies, of Corps projects. The reports notes that such retrospective assessments are integral to sound water planning in general and to adaptive management in particular.
- The need for a greater degree of centralization and streamlining of Corps planning programs and studies. The Corps is highly decentralized, with dozens of district offices spread across the United States. This arrangement may have some limitations with regard to more complex Corps planning studies, and it may inhibit the sharing of information and learning from experience throughout the agency. The reports also observe that Corps planning reports do not always clearly convey key assumptions, methods, costs and benefits, environmental problems and concerns, and conflicts and differences of opinion.
- The Corps and other U.S. federal water resources management agencies today rely on a diverse collection of policies, regulations, and case law that constitute the *de facto* national water policy. Many of these laws have only limited relevance to contemporary water resources needs and in some cases are not fully consistent with more recent laws. This situation occasionally results in confusion (or worse, conflict) between federal agencies. All the study panels discussed these issues, with their recommendations sometimes framed differently. For example, one panel (Analytical Methods) recommended the assignment of interagency coordination responsibilities to a governmental body. Other panels (Adaptive Management, and River Basins and Coastal Systems) called for clarification from the administration and Congress in sorting out inconsistencies within the *de facto* body of national water policy. The Coordinating Committee recommended a slightly different approach in reconciling these types of inconsistencies, calling for the creation of a process to elevate interagency conflicts to higher authority.
- The need to consider implications of study cost sharing (the contribution of a local sponsor to a Corps civil works project). All 216 study panels discussed cost-sharing arrangements for Corps projects, generally noting that increased cost-sharing requirements resulted in a complex mix of positive and negative outcomes. Further investigations into and advice on this topic were beyond the scope and resources of the study panels, but the panels concluded that Congress and the Corps should investigate the full implications of cost-sharing policies.

- The need to create more flexible management processes and strategies. These discussions manifested themselves as comments regarding postconstruction evaluations, adaptive management, and relations with federal agencies and other stakeholders. A recommended overall management strategy, consistent with current budgetary, infrastructure, and social realities, was characterized by the Coordinating Committee report as portfolio planning. The portfolio metaphor refers to the collection of assets that the Corps operates or has responsibility to operate or maintain. These assets include the Corps' physical water management infrastructure—including levees, locks, dams, navigation channels, floodwalls, and ice control structures—and the hydrologic (water) and geomorphic (sediment) resources controlled by that infrastructure.

PEER REVIEW

Congress was particularly interested in the process by which the Corps reviews its planning studies. This subject was granted priority within the "216 studies," with one of the study committees addressing the topic in detail (NRC, 2002). Increasing concerns in recent years regarding environmental impacts, economic evaluations, political pressures, and shifting water management paradigms have led to increased criticism of Corps of Engineers planning studies and projects. The complexity and sophistication of most large water resources planning studies suggest the value and importance of some degree of independent review. There is a strong correlation between the independence of reviewers—in terms of knowledge, association with a project, and organizational affiliation—and the credibility, both real and perceived, of external review. A carefully designed review process for Corps planning studies can increase credibility, improve scientific and technical applications, and help ensure planning studies of the highest quality. The Peer Review Committee report recommends the creation of an internal administrative group within the Corps to coordinate all reviews of Corps planning studies.

Whatever type of review process is implemented within the Corps, the report from the Peer Review Committee recommends that the role of review panels should be to identify, evaluate, explain, and comment on key assumptions that underlie technical, economic, and environmental analysis. Review committees should highlight areas of disagreement and controversies to be resolved by the administration and Congress. Review committees also should be free to comment on topics they deem relevant to decision makers, leaving it to the recipient of the review to decide whether those issues constitute technical or policy issues. Review committees, however, should not be requested to provide a final judgment on whether a particular alternative from a planning study should be implemented.

The Peer Review Committee report recommends that the Corps' more expensive, complex, and controversial planning studies be reviewed by independent, external experts. These independent review teams should not include Corps staff, nor should panelists be selected by the Corps. These independent reviews should be overseen by an organization independent of the Corps. Examples of such independent organizations include professional scientific and engineering societies, the National Academy of Public Administration, the National Research Council, and independent federal oversight groups similar to the Department of Energy's Defense Nuclear Facility Safety Board. Internal reviews are appropriate for less complex and less costly planning studies, and for those that involve lower levels of risk. The Peer Review Committee report also recommends that internal reviews be conducted by review committees that include a balance of Corps staff and external experts.

The Peer Review Committee report recommends that results of a review should be presented to the Chief of Engineers well before a final decision is made on a planning study. Reports from review committees should be public documents and should appear in water resources project planning studies submitted to Congress. The review's primary client—usually the Chief of Engineers—should respond in writing to each key point contained in a review. The chief should either agree with the point and explain how it will be incorporated in the study, or rebut the comment and explain why it is being rejected.

Timing, continuity, and costs of review are key considerations. Corps of Engineers planning studies are conducted in two phases—a reconnaissance phase and a feasibility phase—typically lasting one to two years. The point at which the review should be initiated is not always clear because much depends on a study's complexity and duration. If review is initiated early in the study, however, findings and recommendations can be more easily incorporated into the feasibility study. For more controversial studies the report recommends that reviews are best initiated early in the feasibility phase, or even earlier, during the reconnaissance phase. Multiple reviews conducted at several stages of planning studies may also have value, particularly in more controversial and challenging studies, some of which may require 10 years or more to complete.

ADAPTIVE MANAGEMENT

The traditional focus of the Corps of Engineers' civil works program for water resources has been construction-oriented. This focus has shifted over time, however, as federal budgets for new water projects have declined, as public support for new water projects has waned, and as many of the best construction sites for new projects have been developed. In addition, the Corps, along with many other U.S. economic, business, and management entities, is seeking to augment traditional, capital-intensive, construction-based approaches with strategies aimed at producing greater benefits and flexibility from existing structures and systems. The Corps today finds itself in a setting in which it must pay increasing attention to shifting water project goals, possible operational changes, and the use of feedback and lessons from project outcomes.

As the report from the Adaptive Management Committee points out, the concept of adaptive management gained attention during the late twentieth century as an approach that could help increase natural resources management flexibility and project and system benefits. Adaptive management calls for policies that can be adjusted as new information is gathered and discovered. It calls for the monitoring of outcomes to advance scientific understanding and to help adjust policies or operations within an iterative learning and management process. Adaptive management recognizes inherent variability and dynamics in natural and social systems. It calls for ongoing reassessment of environmental, social, and economic goals in connection with stakeholder collaboration. The true measure of adaptive management is how well it helps meet environmental, social, and economic goals, and the extent to which it increases scientific knowledge and promotes collaboration among stakeholders.

The Adaptive Management Committee report also notes that adaptive management is an evolving concept, and its implementation represents a challenge for a construction- and operations-oriented agency like the Corps of Engineers. Key elements of adaptive management are the establishment of a process for reviewing and revisiting management objectives, a range of management options, monitoring and evaluating outcomes, a framework for incorporating new knowledge (economic, engineering, ecological) into management decisions, and stakeholder collaboration. Adaptive management provides a means of responding to changing conditions through revised management actions, while seeking to avoid costly or irreparable mistakes and unintended consequences. It allows for operational changes that respond to changing social preferences and new scientific information.

The Adaptive Management Committee report also offers some cautionary advice, noting that despite its promise and potential, formal adaptive management practices and programs have not been widely applied. Its successful implementation will thus entail not only patience but also a degree of willingness among stakeholders to find common ground. Stakeholders must agree on at least some fundamentals within adaptive management, such as the key scientific or other questions that they would like to pursue using adaptive strategies. Absent any degree of cooperation, a formal adaptive management program may not be viable.

The report recommends that the Corps implement adaptive management at different scales and in different settings, track progress, and aim to learn from successes and setbacks. There is a spectrum of possible adaptive management approaches.

More passive programs focus on monitoring the results of management actions, while more active programs design specific actions to test multiple models of system behavior. The report recommends that the Corps consider the full spectrum of possible adaptive management approaches, and begin developing guidance regarding suitable approaches in different circumstances. Adaptive management strategies may be particularly useful in large, complex ecosystem restoration projects, which often entail high degrees of risk and uncertainty, along with multiple objectives and phases. The Corps was advised to also promote adaptive strategies based on lessons learned from previous, smaller-scale efforts. Although adaptive management strategies are closely linked with natural resources management projects, they can be used in other systems as well. The Corps should consider ways that adaptive management or similar strategies could be applied to its navigation and flood risk management programs, as well as to ecological restoration. Finally, the report recommends that a Center for Adaptive Management be created within the Corps. This center should start as a carefully planned, modest effort for an initial five-year period and its progress should be periodically reviewed.

Adaptive management programs should systematically incorporate means for stakeholder collaboration into planning and management decisions. The monitoring of physical, biological, and economic aspects of natural systems often poses substantial water resources management challenges. The ambiguities that often attend the monitoring of complex ecosystems can hinder adaptive management's cycle of action, observation, evaluation, learning, and new action. Independent expert review can identify inadequacies in modeling, monitoring, and assessment and can help resolve scientific disputes, and therefore should be part of adaptive management programs.

ANALYTICAL METHODS

The quality and credibility of Corps planning studies have been criticized by some groups (e.g., NWF and Taxpayers for Common Sense, 2004). If these criticisms are valid, explanations for the problems could include limited resources for the development and applications of sophisticated analytical methods and models, increasing competition for engineering talent from the private sector, and a lack of clarity of planning objectives and policy direction. The 216 Analytical Methods Committee report (NRC, 2004d) reviewed Corps planning procedures as embodied within the federal *Principles and Guidelines for Water and Related Land Resources Implementation Studies ("Principles and Guidelines,"* or simply, *P&G*; WRC, 1983) and within the Corps' own *Planning Guidance Notebook* (USACE, 2000).

The Analytical Methods Committee report finds that the Corps is hindered in its ability to define clear management directives because of inconsistencies in the large body of *de facto* national water policy that guides the agency. To provide clearer direction to the Corps the report recommended that the administration and the Congress, in cooperation with the states, should reconcile inconsistencies within this "policy." The demise of the federal Water Resources Council in the early 1980s resulted in the loss of a key forum for interagency collaboration on water policy and management issues. As a result, administration-level coordination has been much less frequent, and today there is an abundance of interagency conflicts and loose ends. The report thus also recommends that a government entity be charged with coordinating water policies and activities among the administration, the Congress, the states, and federal agencies with water resources management responsibilities.

The water resources programs of the Corps (along with three other federal agencies) fall within the scope of the planning guidelines set forth in the federal *Principles and Guidelines.* This document, issued by the former federal Water Resources Council, has not been updated for over 20 years. Over this period there have been many changes and advances in planning and analytical techniques, such as valuation techniques, adaptive management, and shifting views of stakeholder participation. The report thus recommends that the federal *Principles and Guidelines* be revised to better reflect contemporary management paradigms, analytical methods, legislative directives, and social, economic, and political realities. Regardless of whether the administration chooses to revise the *Principles and Guidelines,* the report recommended that the Corps draft a revision to its *Planning Guidance Notebook* that is consistent with the Analytical Methods report and present it to the administration.

When the Flood Control Act of 1936 was signed into law, conventional wisdom dictated that a proposed water resources project would be considered viable only if its projected benefits exceeded the projected costs. Sound benefit-cost analysis is still recognized as vital to good decision making; however, it is not regarded as the sole criterion regarding public policy or investment decisions, as these analyses may contain substantial uncertainties and may not adequately reflect relevant, difficult–to-measure (often qualitative) factors, such as stakeholder opinions or nonmarket values. The report thus concludes that benefit-cost analysis should not be used as the sole decision criterion in judging whether a proposed water resources planning or management alternative should be approved.

Corps planning studies routinely are hundreds to thousands of pages in length. Crucial assumptions, alternatives considered, models and datasets employed, and other factors are thus often difficult to understand clearly. The Analytical Methods report concludes that a summary document identifying key objectives, primary environmental and social issues, key assumptions and alternatives considered, trade-offs, and benefits and costs (monetized and nonmonetized), would facilitate better understanding among all parties involved in a planning study. This summary should be presented with a consistent format and should be a standard component of all Corps planning studies.

Periodic monitoring of completed projects should be a routine part of project planning and management. Congress should provide resources to conduct retrospective, or *ex post*, evaluations of water projects and systems, as these types of studies are essential to sound water resources planning. These retrospective reviews can serve as effective means for understanding how demands for project services have changed over time or how closely a project has come to meeting its stated goals. The limited number of reviews of Corps projects may represent a missed opportunity to evaluate strengths and weaknesses of planning methods and how project operations have or have not changed to meet changing conditions.

RIVER BASINS AND COASTAL SYSTEMS PLANNING

Water resources project objectives have broadened to include increased emphasis on environmental (e.g., ecosystem health, biodiversity) and social (e.g., risk exposure, economic development, recreation) objectives, which has increased the complexity of water project planning. To meet these demands the Corps is being asked to undertake integrated water project planning, adopt a watershed or regional approach, and include ecosystem perspectives in its planning processes. Integrated water resources planning is endorsed within the scientific and engineering communities and is supported by Corps policy and statements from Corps leaders. Integrated water resource planning at the river basin and coastal system scale provides a framework within which trade-offs among competing objectives can be evaluated; multiple stressors, unintended consequences, and cumulative effects can be identified; and a more complete assessment of the costs and benefits of a project can be examined in a context that incorporates stakeholder interests. Such efforts represent a challenge not only because of the complexity of the contemporary planning environment, but also because of the complex mix of legislation, congressional committee language, administrative rulings, and legal precedent that defines the nation's water policies. The River Basins and Coastal Systems Committee report concludes that the clear policy guidance and consistent funding and authority necessary to support integrated planning at the scale of river basins and coastal systems currently does not exist.

The River Basins and Coastal Systems Committee report also finds that a lack of consistent national policy guidance, together with pressures to quickly develop water projects with well-defined local benefits, has hampered the Corps' ability to consistently plan water resources projects within a broader and integrated systems context. Furthermore, the report notes that efforts to more fully integrate water resources planning across spatial scales must compete with pressures to focus on local projects advocated by local interests and their congressional representatives.

Effective water resources planning requires guidance on evaluating incommensurate objectives and determining the appropriate time and space scales of the study. As noted above, the *P&G* has not been substantially revised for twenty years, and it is weighted toward benefit-cost analyses that are more appropriate for more traditional, single-purpose water resources projects than for complex, multi-objective water and ecosystem restoration projects. Planning guidance should be updated to provide more balanced and complete information on conducting integrated water systems planning within river basins and coastal systems.

Uncertainty is an inherent part of the management of all natural systems. In the face of uncertainty, water resource planning and management require an adaptive approach in which management actions are framed as experiments that are used, in part, to inform and enhance future decisions. In this context it is necessary to identify key elements of the system whose monitoring will indicate the success of the project in meeting its objectives. Consistent monitoring provides the opportunity to change project features in ways that can correct for unintended or inferior results. Ongoing project performance evaluations are important when dealing with increasingly complex and highly interactive systems. The report also concludes that project evaluation should be a routine component of all water project operations, and its costs should be shared with the local sponsor. Because the complexity and potential consequences will vary from project to project, current cost limits on project evaluations should be replaced with a system in which the scope, tasks, standards, and costs of project planning and evaluation are determined on a case-by-case basis within a feasibility study.

An improved water resources planning environment will require the support and cooperation of Congress, the executive branch, and the U.S. citizenry. General policy guidance mandating watershed, regional, and ecosystem analysis is clear and publicly supported by Corps leadership. Political support for true watershed or coastal systems planning, however, has been neither consistent nor strong enough to overcome the challenges of implementing this sophisticated concept. The report thus concludes that changes in planning guidance and Corps institutional procedures can allow for more effective and consistent applications of integrated water resources planning and environmental stewardship in river basins and coastal systems.

COORDINATING COMMITTEE

The Coordinating Committee report notes that the Corps of Engineers today is in a position in which it must maintain and operate an extensive water resources infrastructure that must serve both traditional purposes (e.g., navigation) and a new ecosystem restoration mission. Management of this existing infrastructure of dams, waterways and canals, ports, harbors, levees, and hydropower facilities is important because Corps of Engineers structures control a large portion of hydrologic and geomorphic processes in the nation's major river systems and along its coastlines. Efficient infrastructure management is also important because construction spending on federal water projects has declined over the past few decades, thereby increasing the importance of better management of existing facilities. The Coordinating Committee report thus recommends that the Corps center its planning activities around a concept of "portfolio planning."

The Coordinating Committee report notes that the term "portfolio" is used in Corps planning documents (and elsewhere), and that its use in this setting entails the consideration of all Corps of Engineers assets in managing water and related resources. These assets include (a) physical infrastructure, such as locks, dams, levees, and navigation channels; and (b) the water and sediment resources controlled by that infrastructure (e.g., water and sediments that could be released in dam flows in order to rejuvenate wetlands and floodplains). Portfolio planning includes management of existing infrastructure, the addition of new infrastructure, and—where warranted—removal or decommissioning of infrastructure. It entails evaluation of new investments in the context of existing infrastructure and its operations. As the Coordinating Committee report states, "Portfolio planning does not mean that the Corps program will no longer serve traditional navigation and flood risk management needs, but it does mean that the needs can no longer primarily determine how past project investments are operated and new project investments evaluated" (NRC, 2004f).

The Coordinating Committee report also provides advice on focusing Corps of Engineers program areas and mission. In particular, it recommended that the Corps focus its primary environmental mission on the restoration of hydrologic and geomorphic processes in large river and coastal ecosystems. The Corps of Engineers has made ecosystem restoration a program area on par with its traditional flood and navigation programs. The Coordinating Committee report notes that the Corps' emphasis within the broad field of restoration, however, is currently not well defined and could conceivably cover many different dimensions, including species reintroduction. The recommendation to focus on hydrologic and geomorphic processes was offered for several reasons. One is that hydrology and geomorphology are traditional fields of Corps of Engineers emphasis and expertise. Another is that there are abundant opportunities for the Corps in these realms. Finally, a focus on these components of water management will help delineate the Corps' responsibilities with regard to other federal agencies that are also working in ecosystem restoration (e.g., the U.S. Fish and Wildlife Service and the U.S. Geological Survey).

The Coordinating Committee report recommends that a new study authority be enacted for the Corps of Engineers. It notes that many Corps planning studies today are conducted not for the construction of new infrastructure but to rehabilitate existing infrastructure or to create new operations schemes. The Corps possesses authorities that allow it to conduct planning studies regarding project operations, for example. The two most commonly used continuing authorities are from the 1970 Flood Control Act and the 1986 Water Resources Development Act. The Upper Mississippi River-Illinois Waterway feasibility study was conducted pursuant to the 1970 Flood Control Act. These existing authorities, however, were not designed to help orient the agency's planning processes and priorities toward managing existing infrastructure, and the report concludes that they thus are insufficient for moving the Corps toward the planning portfolio paradigm. The report lists several principles that should be part of the new study authority (NRC, 2004f, p. 6-7).

The portfolio planning concept, a focus on hydrology and geomorphology within ecosystem restoration, and a new planning authority are the first-order recommendations from the Coordinating Committee report. The report also lists several other recommendations that support these first three recommendations. These recommendations are in the areas of planning expertise, resolving interagency differences, regional assessments, use of computer-aided decision making, content of a Chief's Report, reconnaissance-feasibility study distinctions, and backlogged projects that have been authorized but for which funds have not been appropriated.

216 STUDY REPORTS SUMMARY

The U.S. Army Corps of Engineers is one of the nation's oldest and most recognized federal agencies, with a long history of national service. Today, however, as national and global economic, environmental, and trade priorities shift, the Corps is experiencing considerable organizational, technical, and budgetary challenges, a well as challenges to its authority and capabilities. The 216 study committees considered this larger context of Corps of Engineers project planning, and their reports were offered in the spirit of helping the Corps best meet the nation's emerging water resources and related needs.

By recommending procedures aimed at increasing the Corps' decision-making flexibility, the 216 study reports may provide some impetus toward helping the Corps move into a new national water management era. The reports recommend an increased emphasis on postconstruction monitoring and subsequent operational adjustments. This recommendation was made in recognition of the inevitable uncertainties and surprises associated with Corps projects, as well as shifting social preferences for the benefits of civil works projects.

The administration, Congress, and the states need to play more active roles in defining the Corps' missions and programs. This is necessary to coordinate the Corps' efforts with other agencies, to provide clearer direction within a complex and sometimes inconsistent body of *de facto* water policy, to provide adequate resources for the Corps to make necessary transitions and changes, and to forward conflicts that the Corps and other line agencies cannot resolve to higher authority. Finally, there is a need for a greater flexibility of Corps management and planning regimes, which includes an increased ability to monitor postconstruction outcomes and make necessary adjustments. This concept is treated in detail in the report from the Adaptive Management Committee, and also is captured in the Coordinating Committee's portfolio planning metaphor, which describes the broad suite of Corps of Engineers assets—namely, physical infrastructure and the water and sediment resources controlled by that infrastructure.

Few Corps of Engineers planning studies have attracted more attention than the agency's feasibility study for the Upper Mississippi River-Illinois Waterway. The following chapter builds upon the findings and recommendations presented in the 216 study reports, and comments on the implications of those reports for managing the diverse resources across the Upper Mississippi and Illinois River systems.

3

Improving UMR-IWW Resources Planning and Management

The 216 study reports offer numerous recommendations for improving review and planning procedures in Corps of Engineers water resources project planning studies. This chapter builds upon Chapter 2's summary of those recommendations, focusing on concepts from the 216 reports that are especially relevant to UMR-IWW management: interagency coordination and study authorities, study authorizations and legislation, adaptive management on the UMR-IWW, valuation of nonmarket benefits and costs, streamlining Corps planning studies, coordination with the Institute for Water Resources, and peer review.

INTERAGENCY COORDINATION AND STUDY AUTHORITIES

Sectoral and Spatial Coordination

As pointed out in this committee's first two reports, the UMR-IWW has several different uses and they have a variety of impacts on one another. Among the uses of the river (including its current lock and dam structures) and its floodplains are navigation, flood conveyance and storage, outdoor recreation, industrial and municipal water supply, waste assimilation, and maintenance of biological productivity and diversity. Sound management of the UMR-IWW would recognize linkages among different uses and consider trade-offs among management decisions in different sectors. Concerns over the limited degree of integration within the Corps' UMR-IWW feasibility study were expressed in this committee's second report (NRC, 2004b).

The Corps is only one of many water management organizations and users of the UMR-IWW. Effective coordination across water sectors and consideration of cross-sectoral impacts depends upon coordination among such groups as local municipalities, state resources agencies, the Federal Emergency Management Agency, the U.S. Fish and Wildlife Service, and the Corps.

As pointed out in this committee's second report (NRC, 2004b), the notion of comprehensively managing a river basin as a single unit dates back to at least the late nineteenth century, when John Wesley Powell advocated the use of watersheds to define political boundaries in the arid western United States. Many examples could be used to illustrate the importance of recognizing spatial linkages across a watershed or river basin. If levees are not constructed to a similar standard height, for example, one side of a river might be protected at the expense of flooding on the other side (Kelley, 1998). Levees that block off too many backwater areas in upstream locales can contribute to increased flood heights downstream. Reservoir and dam operations in upstream basin locations can affect downstream flows and users, and can result in upstream-downstream and interstate tensions. Urban development, population growth, and agriculture practices directly affect sediment load and movement, and water quality. Prolonged drought conditions across much of the western United States, for example, have contributed to interstate tensions regarding Colorado River dam operations (see Denver Post, 2005).

Planning across the UMR-IWW system should account for these types of linkages across space and across different water uses, both along the river's main channel and across its tributary systems. On the UMR-IWW, water resources agencies ideally will consider not only how various activities affect one another along the mainstem Mississippi River. Furthermore, they should consider how water and land management issues in tributary systems, such as sediment transport and deposition, and sources and transport of nutrients and pollutants, affect water quality in the mainstem Mississippi River and ultimately the Gulf of Mexico.

UMR-IWW Interagency Coordination

As discussed in Chapter 2 the 216 reports call for better coordination among federal agencies, and for better coordination among federal agencies, state and local governments, nongovernmental organizations, and water resources users. The Adaptive Management Committee report, for instance, recommends a strengthening of federal interagency coordination mechanisms to promote adaptive management. The River Basins and Coastal Systems Committee report calls for greater attention to interagency collaboration. The Analytical Methods Committee report recommends that a body be charged to coordinate water resources policies and activities among the Administration, the Congress, the states, and federal agencies with water resources management responsibilities. Finally, the Coordinating Comittee report recommends a process for resolving interagency differences and disputes.

Many aspects of the Corps' restructured UMR-IWW feasibility study reflected concerted efforts aimed toward better interagency coordination and collaboration. At the policy level the Corps was instrumental in forming a federal interagency Principals Group that included senior-level representation from the Corps, the Department of Agriculture, the Department of Transportation, the Environmental Protection Agency, and the Fish and Wildlife Service. The Principals Group was created in 2001 and periodically convened representatives from these agencies to ensure that significant issues were addressed in an interagency setting and that the Corps' evolving study had the support of other agencies.

Within the feasibility study the Corps convened over 30 governmental and nongovernmental organizations in the Upper Mississippi River region to discuss the study's overall direction and to address specific issues as they arose in the course of the study. A stakeholder group—including other federal agencies, state resources agencies, and nongovernmental organizations—actively participated in the design of ecosystem restoration plans. The level of regional participation in the feasibility study is noteworthy. The Corps also conducted an ambitious campaign to promote collaboration among federal, state, and local agencies, and to promote public involvement in the development of alternatives.

The Principals Group itself may have been beneficial, but this committee's investigations focused largely on the study's analytical components and did not include examination of the structure or functions of the Principals Group. Interagency collaboration will continue to be important on the UMR-IWW, as well as in other large river and water systems in which the Corps is working (e.g., Florida Everglades, Missouri River). Current plans call for the Principals Group to be retained and to continue promoting interagency collaboration on the UMR-IWW.

Given the importance and prominence of the UMR-IWW Principals Group, an independent accounting of the group's achievements, challenges, and future prospects would be relevant and useful. For example, if there were functions of the Principals Group that were particularly valuable or that could be improved, these could be applied to future UMR-IWW management decisions and processes. A clear understanding of the Principals Group charter and mandate, the extent to which this mandate was successfully executed, and the key challenges encountered in trying to realize that mandate, could also be of value. It would be instructive to know specific questions and tasks (if any) that were presented to the group, what data and information were provided by the Corps, what decisions were reached, and the extent to which they were followed. It would be useful to know how the Principals Group facilitated cooperation that might not have otherwise occurred, and whether the feasibility study benefited in any demonstrable way as a conesquence of the workings of the Principals Group. It also would be interesting to learn to what extent the Principals Group discussed the concept of adaptive management, and whether the group discussed means by which adaptive principles could be employed in managing UMR-IWW resources.

Groups with similar mandates and structures to the UMR-IWW Principals Group have been established in several other river and aquatic ecosystems across the United States, such as California's CALFED program, the Glen Canyon Dam Adaptive Management Program, the Louisiana Coastal Area Ecosystem Restoration Project, and the South Florida Ecosystem Restoration Task Force. Although these efforts are seen by many participants and observers as useful, these views (which are not universal) are nearly always based on anecdote. A formal, independent review of the UMR-IWW Principals Group experience should be useful to other U.S. federal interagency initiatives for managing rivers and water resources systems, or even regions of the world with international river commissions or similar arrangements. A review of the UMR-IWW Principals Group should thus consider other U.S. interagency water management forums, and how similarities and differences between these various groups affect decision making and river management outcomes.

The Corps of Engineers should enlist the services of an independent investigator or a small group of investigators to review and assess the experience with the federal, interagency Principals Group for the UMR-IWW feasibility study. The investigator report should also consider experiences with other high-level interagency groups that have been assembled to help manage large U.S. river and aquatic systems.

Clarification of Authorities, Policies, and Other Legislation

Several of the 216 study reports summarized in Chapter 2 identified problems associated with the existence of a large number of policies, acts, authorities, and other directives related to water resources management that are not fully consistent with one another. These 216 panel reports thus issued calls, for example, to Congress and the administration to help clarify inconsistencies or to more carefully define priorities within this large body of national water "policy." On the UMR-IWW, examples of this large body of legislation are the 1930 Rivers and Harbors Act that authorized the 9-foot channel project on the Upper Mississippi River, numerous Water Resources Development and Flood Control Acts that authorized various water- and flood-related projects, the Endangered Species Act of 1973, and the Upper Mississippi River Management Act of 1986.

As pointed out in this committee's second report (NRC, 2004b), these multiple directives for Upper Mississippi and Illinois River operations are not always fully consistent with one another. This often requires the Corps to choose as to which authorization(s)—and therefore which group of river resource users—are to receive priority. Multiple acts and authorizations also have different implications for channel depths and river flows. For example, an authorization to maintain a minimum 9-foot channel is clear enough; but other authorized purposes, such as protection of endangered species habitat or improvements in river system ecology, may suggest different river channel depths or river flows in different seasons. Within this muddled legislative setting the Corps generally interprets the 1930 authorization for the 9-foot channel as the overriding authority in managing the UMR-IWW system.

Although some may argue that this type of policy ambiguity provides flexibility for the administration and the Congress in dealing with multiple constituencies in the UMR-IWW, the existence of conflicting directives places the Corps of Engineers—an executive-level line agency—in the uncomfortable position of choosing which constituency is to receive priority. Moreover, the primacy that the Corps accords to the 9-foot channel in UMR-IWW management decisions effectively rules out several potential trade-offs (e.g., maintaining an 8-foot channel at some times of the year) between commercial navigation and other related uses, such as boating and commercial and recreational fishing.

Another example of federal direction that should be revised and clarified is within the federal *Principles and Guidelines* (*P&G*), which has been unchanged since 1983. This 22-year-old document is regarded by many as the conceptual basis of U.S. federal water resources planning studies, yet it is silent on the subject of ecosystem restoration. The Corps adopted a National Ecosystem Restoration account in its 2000 planning guidance (USACE, 2000) as a legitimate project purpose and objective, yet the *P&G* continues to support single-purpose project planning dedicated to the maximization of National Economic Development. The report from the 216 study panel on analytical methods notes this and other shortcomings of the *P&G*, leading to a recommendation in that report that the *P&G* be revised (NRC, 2004d).

To help the Corps of Engineers and other federal and state agencies better manage and understand the federal intent for use of UMR-IWW resources, the administration and the Congress should clarify relative priorities among the multiple laws, executive branch guidances, and congressional reports that govern UMR-IWW management.

ADAPTIVE MANAGEMENT

The Corps of Engineers made strong efforts to include adaptive management as an element within the UMR-IWW feasibility study. As described in the 216 panel report on adaptive management, there are less and more formal versions of the concept (e.g., passive and active adaptive management). There does not appear to be any adaptive management program that represents a global prototype. Nevertheless, several themes and components are widely seen as important to adaptive management. In its most succinct form adaptive management can be described as a process of learning while doing (Lee, 1999). An ideal adaptive approach would accept that management actions are impermanent and represent opportunities to learn more about the system at hand. Results from those management actions would be carefully monitored and evaluated, and then used to inform and adjust future actions.

Input from both scientists and stakeholders is increasingly viewed as essential to natural resources planning, particularly in situations characterized by conflict. This implies that some steps need to be closely followed in establishing a viable adaptive management program. Stakeholders (including water managers) should first agree on the key scientific and related water management questions to be pursued in connection with operating a project such as the UMR-IWW. Those agreed-upon questions then should be used to organize subsequent monitoring and evaluation efforts. Such prior agreement on questions that all interests want to see pursued is important to the independence of the scientific and management staff managing the learning and adaptive processes. The answers they develop may not be the ones preferred by all interests, but if the list of questions has been agreed to earlier, this will promote the independence (and perhaps a degree of protection) of the staff charged with evaluation and monitoring. When this process can persevere long enough for the project to perform under a wide range of natural conditions and human and social stresses, the result is the iterative learning cycle promoted by adaptive management.

Points listed in the 216 Adaptive Management Committee report that may have UMR-IWW applications are:

- Adaptive management practices can be useful across a variety of scales and settings.
- Although adaptive management has clear implications for managing ecosystems, it can be used to manage other types of systems, such as transportation infrastructure.
- Adaptive management does not represent a panacea for water resources management yet it holds great promise for helping the Corps to better accommodate shifting social preferences and new scientific knowledge.
- Adaptive management is not an end in itself, and its value ultimately will be measured by its ability to meet environmental, social, and economic goals and to enhance scientific knowledge.

- A viable adaptive management process requires some level of agreement among participants; if there is no flexibility or willingness to compromise among stakeholders, the ability to manage adaptively will be sharply limited.
- Adaptive management can be used as an approach in individual projects, but it also represents a broader management perspective.

The Corps has accumulated a wealth of experience in UMR-IWW system management over the years, and that record of experience should be reviewed as part of promoting more adaptive regimes in future UMR-IWW management. The Corps employs several elements of an adaptive approach in operating and maintaining the existing navigation infrastructure. Dam operations, for example, involve frequent measurements of water levels, weather forecasts, and traffic levels to make appropriate gate adjustments in response to this information. Maintenance and rehabilitation of the structures is based on periodic inspections. The Corps has conducted experimental drawdowns of navigation pools at Pool 8 near LaCrosse, Wisconsin. The Corps also has been a cosponsor of the federal-state Environmental Management Program (EMP), which includes a Habitat Rehabilitation Program (representing about two-thirds of EMP funding) and the Long Term Resource Monitoring Program (representing about one-third of EMP funding). Through these programs the Corps has gained experience in monitoring ecological changes and human impacts along the Upper Mississippi River. The value of ecosystem monitoring programs such as the EMP can be enhanced by ensuring good linkages between monitoring results and resource management decisions. These linkages can be strengthened through a process in which stakeholders and managers identify key questions and topics to be pursued through a monitoring program.

The UMR-IWW is a large interstate river system with numerous management agencies and users. Although such large, complex ecosystems could especially benefit from adaptive approaches, the size of such systems and the multiple managers, users, and points of views across the region pose challenges to implementing large-scale, multiple-stakeholder adaptive management. In such settings it may thus be useful to identify subsystems within the UMR-IWW of specific interest, such as the Pool 8 area mentioned above, in which managers and stakeholders can agree to learn from their experience over time. It should also be noted that many actions and factors that affect UMR-IWW management go beyond Corps of Engineers authorities and programs. For example, additional urban development in floodplain areas throughout the basin removes natural floodwater storage, which may contribute to an overall reduction in the river system's flexibility and resilience to cope with floods. Nevertheless, many actions could be taken—by the Corps and by others—to help improve knowledge of UMR-IWW water-related sectors and to enhance management flexibility and social benefits.

Although the UMR-IWW feasibility study contemplated adaptive management for both navigation improvements and ecosystem restoration, the report particularly highlighted the application of adaptive principles to the ecosystem restoration component. Ecosystem management is amenable to many of the principles articulated in the adaptive management literature; indeed, systems ecologists in the late 1970s generally are credited with formalizing adaptive management principles (see Holling, 1978). Adaptive principles have applications beyond river ecology, however, and examples of adaptive actions contemplated—but not necessarily implemented to date—by the Corps are:

- Locks and dams can be extended in a phased process, with ongoing reassessment of demand for navigation and effectiveness of both nonstructural and structural means of reducing congestion.
- Nonstructural means for better managing existing systems—such as waterway traffic management systems—can improve management efficiencies and squeeze more benefits from existing infrastructure, as well as enhance navigation system learning and adaptability.
- An adaptive approach facilitates comparisons of models and forecasts with actual outcomes, as well as other types of *ex post* studies. For example, alternative locking rules generated by model predictions and consultation with navigation industry can be implemented to test how well they work in reality. Similarly, grain export forecasts can be compared with actual export levels to see how well they compare with real-world outcomes, with those comparisons being used in an iterative process to help inform and improve future forecasts.

The 216 Analytical Methods Committee report recommends that "periodic reviews of completed projects should be a routine part of Corps water project planning and management" (NRC, 2004d). The 216 Adaptive Management Committee report offers a similar recommendation: "Post-construction evaluations should be a standard for adaptive management of Corps projects and systems" (NRC, 2004c). Despite the importance of these types of retrospective investigations, they have historically not been a planning and design standard in water resources management. As national and global water expert Gilbert White pointed out years ago, "We could fill a large room with documents drawing up what are considered the best plans for an analysis of problems in river basins around the world . . . On the other hand, the literature about what has happened after any of the projects have been carried out can be assembled on one end of a small table. There is no tradition of making retrospective or evaluative studies of the consequences" (White, 1971). Successful implementation of adaptive management may rely upon effective *ex post* evaluations (Jacobs, 2002).

The Corps thus should consider ways to apply adaptive management principles to its entire UMR-IWW portfolio. As defined in the report from the 216 study

Coordinating Committee, this portfolio includes both physical infrastructure (e.g., locks, dams, and levees) and water and sediment resources controlled by that infrastructure. Effective adaptive approaches will learn from past outcomes and decisions through reflective, *ex post* evaluations. Comparisons of past forecasts with actual outcomes, and comparisons of anticipated project outcomes with actual outcomes, will enhance future modeling efforts and planning decisions.

In moving forward with UMR-IWW adaptive management actions, it should be recognized that adaptive management is not a project add-on to be implemented, limited, or set aside according to budgetary constraints. Rather, it is a process and perspective that should become part of the organizational fabric. The administration and the Congress should support the Corps in its efforts to integrate adaptive concepts into the operations of its entire UMR-IWW portfolio, including ecosystem restoration projects, transportation infrastructure, and waterway traffic management. Retrospective comparisons and studies can improve future forecasts and other aspects of UMR-IWW decision making and should be seen as integral to an adaptive approach.

VALUATION OF NONMARKET BENEFITS AND COSTS

Background

Large infrastructure projects commonly involve external costs that are not reflected in market transactions. Typically these costs result from environmental impacts of the project, such as nonmarket, external economic costs imposed on users of environmental goods and services. Many public sector projects, such as most water resource and all ecological restoration projects, are distinguished by large nonmarket benefits as well as nonmarket costs. Failure to fully account for these nonmarket effects seriously distorts the evaluation of projects. Where nonmarket benefits predominate, for example, the result may be the underprovision of such benefits and the diversion of public moneys to projects that are ultimately less beneficial.

Prior to 1983, the federal *Principles and Standards* (or *P&S*) that governed water resource project evaluation had two coequal objectives: (1) National Economic Development (NED) and (2) Environmental Quality (EQ) (WRC, 1973). The design of the *P&S* was intended to promote full consideration of nonmarket environmental effects, both negative and positive, by means of a multicriteria decision-making approach. However, in 1983, the *P&S* document was rescinded and replaced with the *Principles and Guidelines* (*P&G*), which eliminated the EQ objective but retained the EQ account creating what is, in effect, a single objective with an EQ constraint (WRC, 1983). This approach may limit negative external effects, but it does not necessarily recognize environmental improvements (such as ecosystem restoration) that may result from project implementation.

If nonmarket benefits provided by a project can be valued in monetary terms, they can be included in a project's NED account. Methods for accomplishing this were well known in 1983, and the *P&G* document discusses the use of a variety of valuation methods for nonmarket goods, even providing an appendix describing contingent valuation methods (WRC, 1983). Furthermore, the Corps and other agencies had extensive experience with valuing the nonmarket benefits of conventional water supply projects, such as flood damage reduction and some kinds of recreational benefits. For these reasons one might have expected increased interest in valuing nonmarket benefits after 1983, but there is no evidence of such a trend. The report from the NRC 216 study panel on Analytical Methods notes that the "Corps was under little pressure from the administration to develop techniques for monetization of environmental goods and services" (NRC, 2004d, p. 61). Accordingly, little has been done to expand the valuation of nonmarket effects of Corps projects. This is true despite rapid progress in the evolution and application of such methods elsewhere.

The Corps' primary civil works policy research unit is its Institute for Water Resources (IWR). Located at Fort Belvoir, Virginia, the IWR's staff conducts research on a variety of economic, engineering, and environmental topics that pertain to Corps project operations and planning. The IWR has allocated substantial resources to developing methods for quantifying and valuing restoration benefits, often with little apparent impact on practice. This remains true despite major initiatives by the Corps in the area of ecosystem restoration, where the need for valuing nonmarket benefits seems large. A 2005 NRC report that considered the application of nonmarket valuation methods to ecosystem services concludes that "the current state of both ecological and economic analysis and modeling in many cases allows for estimation of the values people place on changes in ecosystem services" (NRC, 2005, p. 242). This section addresses the need for more extensive valuation of nonmarket effects of Corps projects, as well as various methods that are used for this purpose.

Valuation Methods for Ecosystem Restoration Benefits

Despite thorough understanding and extensive experience with a range of nonmarket valuation techniques, the application of these methods to ecosystem restoration has been controversial. In his seminal 1993 treatise on valuation methods A. Myrick Freeman expressed doubt that these techniques could be applied to "such things as biodiversity, the reduction of ecological risks, and the protection of basic ecosystem function . . . except where nonuse values are involved or where people use ecosystems" (Freeman, 1993, p. 485). This concern was echoed in a more recent Corps of Engineers report, which pointed to "considerable technical obstacles [that] stand in the way of comprehensive monetary accounting of restoration project benefits" (Stakhiv et al., 2003, p. 129). The Stakhiv report notes both scientific obstacles ("in tracing the links between restoration actions and service outcomes underlying all possible routes to human benefits") and economic obstacles (where benefits "have no close connection to marketed goods").

The 2005 NRC report on the valuation of ecosystem services discusses the meaning of value for ecosystem services, noting a number of different dichotomous views: instrumental versus intrinsic, anthropocentric versus biocentric, and utilitarian versus deontological (NRC, 2005). These differing perspectives lead many to view ecosystem value as a complex, multidimensional property, not readily amenable to quantification or economic valuation. To perform such a valuation implies an anthropocentric, utilitarian concept of value that may not capture all relevant aspects of value, particularly for such things as "biodiversity, the reduction of ecological risks, and the protection of basic ecosystem function," as noted by Freeman. The report notes that there are many situations in which even a partial or one-dimensional assessment of value is useful or necessary. In such cases economic valuation approaches can be employed.

The 2005 NRC report also reviews approaches to the economic valuation of ecosystem services, including careful analysis of available methods and various objecttions to their use (NRC, 2005). Particular attention is given to tracing the connections between ecosystem structure, functions, and the resulting ecosystem services. It should be noted that it is the flow of ecosystem services that has potential economic value to humans. Ecosystem services are those consequences of ecosystem functions that have the ability to enhance human well-being. Accordingly, people can be expected to express a willingness to pay for these services; that is, these services have an economic value. The 2005 NRC report concludes that although relationships between ecosystem structure and functions are not fully understood, much is known. It further concludes that there is less knowledge of the relationships between ecosystem functions and services. There is, however, considerable knowledge of the relationships between ecosystem services and economic value.

In examining valuation methods potentially applicable to ecosystem restoration outputs, the NRC report adopts the usual distinction between revealed preference methods (values are imputed from observed market behavior involving complementary goods) and stated preference methods (values are expressed directly by responses to a survey). Revealed preference methods include, for example, analysis of averting behavior, the travel-cost method, and hedonic price analysis. Commonly used stated preference methods include contingent valuation, contingent referendum, and conjoint analysis. The report also discusses some cutting-edge methods, such as combined revealed-stated preference methods and the use of computable general equilibrium models of ecologic-economic systems.

After considering objections, limitations, and capabilities, the report concludes:

> Nonetheless, the current state of both ecological and economic analysis and modeling in many cases allows for estimation of the values people place on changes in ecosystem services, particularly when focused on a single service or a small subset of total services. Use of the (imperfect) information about these values is preferable to not incorporating any information about ecosystem values into decision-making (i.e., ignoring them), since the latter effectively assigns a value of zero to all ecosystem services (NRC, 2005, p. 242).

A review of the literature on this subject makes clear that reliable valuation of the totality of ecosystem services is well beyond present capabilities. This is true because the kind of anthropocentric, utilitarian approach implied by the focus on ecosystem services may not capture everything that is regarded as valuable. Further, current valuation methods may not be available for every ecosystem service in every situation. But it is equally clear that many ecosystem services that have use or nonuse value to humans can be credibly valued in monetary terms, given adequate understanding of the linkages from ecosystem structure to function, and function to ecosystem service. The 2005 NRC report notes the limitations to our understanding of these connections but also argues that much is known and that what is known is a sufficient basis for valuation in many cases. A frontier for research is the proper quantification of the links between improvements in such parameters as connectivity, nutrient cycling, critical habitat, and biodiversity and the resulting use and enjoyment humans derive from such improvements.

Current Status of Valuation Methods

When projects have significant benefits or costs that have not been expressed in monetary units, the ability of benefit-cost analysis to distinguish between projects that are beneficial and those that are not is limited to several specific situations. Otherwise, attempts to use benefit-cost analysis to justify projects, rank alternatives, or allocate project funds will lead to serious distortions. In general, projects with large net monetary benefits (conventional water resource projects) will be favored over those with large nonmonetary benefits (ecosystem restoration projects), regardless of their ultimate value to society. Even analyses of conventional water resources projects have generally failed to value all project effects, such as all environmental costs associated with the construction of a dam or channelization of a stream.

Well-understood methods exist for credibly valuing some benefits of ecosystem restoration in monetary units. These methods are not typically applied by the Corps of Engineers. As scientific knowledge accumulates and, in particular, as more is learned about linkages between ecosystem functions and services, opportunities for valuation of ecosystem services will expand.

The current state of ecosystem science and economic analysis clearly supports valuation of the benefits of ecosystem restoration. In some cases the valuation of these benefits can be as complete as the benefits derived in more traditional projects, such as flood control and navigation projects. Even the valuation of some ecosystem restoration benefits will improve the quality of decision making for these projects. In many cases valuation of the most obvious benefits will be sufficient to demonstrate feasibility. In other cases even incomplete valuation may allow for a credible comparison of the remaining nonmonetary benefits against net monetary costs. In either case there is no reason to continue restricting applications of nonmarket methods to traditional categories. All Corps water project benefits and costs should be valued in monetary terms to the extent possible.

STREAMLINING CORPS PLANNING STUDIES

The 216 report from the panel on analytical methods noted that "Corps of Engineers planning studies, and attending appendices and other documents, are often hundreds of pages in length. This quantity of information often makes it difficult to identify and comprehend all important assumptions, alternatives, models employed, data sets, and other factors" (NRC, 2004d, p. 8). It was recommended that the Corps develop a standard summary document with a consistent format across all studies that would identify "key environmental and social issues, primary assumptions, alternatives considered and evaluated, objectives sought, benefits and costs (monetized and nonmonetized), trade-offs, and stakeholder perspectives and differences" (NRC, 2004d).

The volume of material presented in various drafts of the UMR-IWW feasibility study, measured in thousands of pages, made the report particularly difficult to comprehend. These types of federal resources planning documents are written to address all relevant statutory (e.g., National Environmental Policy Act) and other guidances. As a result they tend to be cumbersome, difficult to organize, and often confusing to the reader. Many important topics that were presented in appendices were not well integrated into the main report. For example, the Corps indicated verbally that its plans for restoration included navigation pool drawdown experiments (e.g., varying elevation of pool water levels). Unfortunately, most of the information about these experiments was contained in appendices and not noted in the main report. Although the report contained an executive summary, the summary was difficult to follow and did not highlight and clearly present the key issues discussed in the report. As suggested by the report from the 216 panel on analytical methods, decision makers and analysts would have benefited from a summary document.

Some of these organizational and presentation problems stem from attempts to merge a complex planning study with complex environmental impact assessment. The statute-based environmental impact assessment should not be confused with the environmental analyses that are integral to a water resources planning study, and that should be addressed therein. This committee, for example, has encouraged the Corps to better integrate economic and environmental issues into the feasibility study (NRC, 2004b). Considering the interplay between economic and environmental (and engineering and social) issues within a feasibility study, however, is a process separate from the conduct of an environmental impact statement. The former is part of a sound water resources planning study, while the latter is conducted pursuant to federal statute to determine the environmental impacts of a proposed federal action. The latter is also conducted, to a large degree, in a process that is distinct from the feasibility (planning) study.

Other problems within the study arose from efforts to address and provide information about all potentially controversial topics in the main body of the report. The scope of the study was not clearly delineated, with the result that much peripheral information was included that was not directly relevant to the study. Not only did this make the study difficult for readers to comprehend, the size and organization of the study surely made it difficult for Corps staff to edit and update through the course of the study.

To streamline the preparation of future complex Corps planning reports like the UMR-IWW feasibility study, to enhance their presentation, and to improve their readability, the following steps should be taken:

- **There should be a succinct and substantive summary of the key planning issues addressed in the report (this parallels a recommendation from the 216 studies Analytical Methods report; see NRC [2004d]).**
- **Although it is essential to integrate economic and environmental issues into water project plans, the practice of merging water project planning reports with environmental impact assessments should be reconsidered. If these reports are to be merged, the process of integrating them should focus on presenting a clear understanding of the overall report, as well as linkages among its main components. Each of these types of reports presents considerable preparation and presentation complications, however, and in large, complex planning studies, the separation of the envi-**

ronmental impact assessment from the feasibility and project planning studies should be considered.

- **For studies of this magnitude, a full-time staff— technical editor(s) and technical writer(s)—should be retained to oversee the report production and presentation process, including the display of Web-based documents.**
- **Technical details of planning studies should not be included in the report's main body but should be included in appendices, both on paper and in Web-based documents.**

COORDINATION WITH THE INSTITUTE FOR WATER RESOURCES

The Corps of Engineers operates in a planning environment that is increasingly complicated and that poses stringent challenges to the agency's analytical capabilities. In its more complicated planning studies, the Corps often is required to provide credible, science-based forecasts of future conditions (e.g., waterway traffic levels); to create and apply sophisticated, large-scale economic models; or design a program for large-scale ecosystem restoration. These challenges require greater staffing and organizational capabilities than in previous eras.

The Corps conducts water resources planning studies in a variety of settings and on a breadth of topics. These studies run the gamut from those conducted for small areas and for primarily a single purpose (e.g., the raising of a levee for a community or restoration of a backwater wetland area) to those conducted across multiple states for multiple purposes. These latter types of studies, which include the UMR-IWW feasibility study, pose greater analytical challenges and are more closely monitored by multiple interest groups. The Corps should consider ways to anticipate and prepare for future large-scale studies similar to the UMR-IWW feasibility study.

There are various means by which the Corps could strengthen its planning capabilities. Some steps for improvement were offered in recommendations in the 216 study reports. Another step the Corps could take to improve its planning capacity is to periodically designate a planning study (or a given portion of a planning study) as one of special interest. Studies could be designated as being of special interest when they represent significant technical or organizational challenges that the Corps can expect to face more often in the future. Many aspects of the UMR-IWW could fit into this description, including projections of future traffic levels, use of nonstructural means for managing waterway traffic, valuation of ecosystem benefits, identifying measures to achieve optimal levels of ecosystem restoration, and public participation. This section discusses organizational and technical aspects that the Corps should consider in conducting its more complicated and challenging planning studies.

Planning studies for Corps of Engineers water resources projects typically are conducted within one of the Corps' 41 district offices. There are instances, however, in which planning studies are conducted across multiple Corps districts; this was the case with the UMR-IWW study, which employed staff from several district offices within the Corps' Mississippi River Valley Division. Guidance also was provided by staff from Corps Headquarters in Washington, with the assistance of the Corps' Institute for Water Resources. Deployment of staff from across the agency is consistent with a recommendation from the 216 study Coordinating Committee, which recommended the creation of specially chartered study teams for more complex and controversial Corps planning studies (NRC, 2004f).

In connection with the UMR-IWW feasibility study the Institute for Water Resources (IWR) created a National Economics Technologies (NETS) Program, which aims to develop economics-based models and techniques to facilitate studies of waterway traffic flows and management. The IWR conducts research on a variety of economic, engineering, and environmental topics that pertain to Corps project operations and planning. Staff from the IWR explained the elements of its NETS program to this committee. The NETS program is consistent with recommendations in this committee's 2004 reports regarding river traffic forecasts, navigation benefits estimation, and consideration of nonstructural measures to improve waterway traffic management. The Corps deserves credit for allocating resources to this potentially valuable research; the NETS program holds promise for helping reduce waterway congestion and thus increasing overall benefits from UMR-IWW system operations.

To ensure that IWR research is relevant to and being appropriately applied within Corps planning studies, it is important that (1) research addresses problems relevant to current planning; (2) research results are credible and usable; and (3) field-level planners are aware of those results. At the same time, the IWR is obliged to anticipate future analytical challenges, even those not yet identified by field staff. IWR is expected to develop methods and techniques that reflect the state of the art, not merely answers good enough to address today's (or yesterday's) problems. This implies a trade-off between the aspirations of researchers and the needs of planners. It also implies effective communication between the field and the researchers.

There are various ways that communication and dissemination of ideas could be accomplished. The main requirements are that IWR researchers should be presented with key technical problems arising in the Corps district offices, and that IWR research results be vetted and provided to district-level planners in an understandable and usable form. The issues may be technical, operational, economic, or ecological. The determination of which issues can be researched, and research priorities, must be made by IWR, although with comment and review by field staff. Perhaps the most difficult task is packaging and delivering IWR research results in a way that field planners can make effective use of new models and analytical methods. The NETS program presently addresses these objectives through periodic meetings with the Inland Waterways Center of Expertise group, and the Ports Center of Expertise group, augmented by newsletters and briefings.

The Corps would benefit from communications with other federal agencies regarding methods for disseminating research results and for developing peer review procedures. For example, the EPA uses its "812" studies for developing benefit-cost analysis tools that are used throughout EPA (Section 812 of the Clean Air Act Amendments calls for the EPA to produce an analysis of the costs and benefits of the Clean Air Act from 1970 to 1990). The EPA has a special advisory committee for 812 studies to ensure that up-to-date research results are incorporated in the studies. EPA's Science Advisory Board also reviews 812 studies to ensure that they are scientifically sound. EPA specifically allocates staff and resources to develop the tools for these studies to ensure that they are credible and analytically sound and can be used in other EPA programs. The Corps should look to this example and relevant experiences in other agencies to help ensure that contemporary and credible research theories and methods are being employed within Corps planning studies.

PEER REVIEW

Reviews of different portions of the Corps' UMR-IWW feasibility study were conducted through a variety of means, including independent review by two National Research Council (NRC) committees. The first, Phase I, NRC committee was convened in 2000 and issued its report in 2001. The Phase II committee was formed in mid-2003, at a point during which the restructured feasibility study process was underway. This committee thus was placed in a position of trying to keep pace with ongoing Corps actions while simultaneously sifting through background materials developed over the prior 15 years.

The 216 Peer Review Committee report concludes that reviews generally will be of greater value when initiated earlier in the planning process. Although this conclusion may often hold true, when reviews are started early in the process, reviewers may be requested to keep pace with an evolving study or set of studies. Reviews that are initiated in the early stages of a long-term planning study like the UMR-IWW feasibility study may entail challenges in attracting qualified reviewers who are willing to follow a decades-long planning study over its entire course. The Peer Review Committee report recognizes some of the problems with long-term reviews, noting the following:

> The Corps' most challenging planning studies . . . may require over 10 years to complete. At the same time, it is important that panelists focus on their review of the planning study, and not

> become defenders of their recommendations. To guard against this—especially in lengthy planning studies—different review panels may need to be appointed at different stages of the study.

To the Corps' credit, many steps recommended in the 216 Peer Review Committee report regarding independent review were followed during the UMR-IWW study (the Corps has also recently issued an Engineering Circular on the Peer Review of Decision Documents; USACE, 2005). For example, the Corps prepared documents explaining its agreement with certain recommendations in this committee's earlier reports and how the Corps intended to use them, or why they chose to disagree with a recommendation. It is worth pointing out that the NRC serves as an adviser to the Corps and to its other study sponsors, and the Corps is thus free to accept or reject any comments from an NRC committee. During the review process, the Corps was cooperative and open in sharing information, participating in discussions with the committee, and hosting committee visits on the Upper Mississippi River. The Corps also helped ensure that the committee spent time with interest groups and nongovernmental organizations that were not fully satisfied with portions of the feasibility study.

The reviews of the NRC committees raised several issues and areas for improvement within the feasibility study. Short of an explicit evaluation of how the Corps was able to incorporate NRC committee advice into its feasibility study (an evaluation beyond this committee's scope), it is difficult to say precisely how independent review may have strengthened the results of the study.

4
Epilogue

Efforts by the U.S. Army Corps of Engineers to improve the Upper Mississippi River for commercial navigation date back to the early 1830s. The initial efforts involved the removal of the numerous snags that filled the river at the time. Other nineteenth-century navigation improvement programs included dredging of navigation channels, blasting of rapids, construction of dikes, jetties, and wing dams, and initiation of a 4½-foot channel project in 1878. Even during this era, some observers sensed the enormity of the task that the Corps had embarked upon; the Corps' efforts led Mark Twain to quip that "the military engineers of the [Mississippi River] Commission have taken upon their shoulders the job of making the Mississippi over again—a job transcended in size by only the original job of creating it" (Twain, 1883, pp. 301-302).

But the Corps' efforts to improve commercial navigability continued: In 1907 Congress authorized a 6-foot channel project, and in 1930 a 9-foot channel was authorized for the Upper Mississippi River. The 9-foot channel project, constructed largely during the 1930s, created a series of low-head dams, locks, and navigation pools on the Upper Mississippi. On the Illinois Waterway, construction began earlier, with the first lock and dam completed in 1871 (by the State of Illinois). The 9-foot channel project initiated several lasting, large-scale ecological changes to the system that continue to affect the river ecosystem and its users today. In addition to the navigation project, several other anthropogenic changes have affected river ecology and water quality, including levee construction, construction of hydropower dams, floodplain and watershed agricultural practices, pollution loads, deforestation, and population and urbanization trends.

A key lesson from the past 175 years of managing the Upper Mississippi and its resources is that it has never been possible, in Twain's words, to "fetter and handcuff that river and boss him" (Twain, 1883, p. 302). Over that period an implicit goal of Corps river management policy and practices has been to manage the river in a way that maximized overall benefits to society. As part of the 1936 Flood Control Act, the Corps was mandated to ensure that benefits from proposed flood control projects exceeded the costs; this criterion was extended to all water resources projects by 1950. But as Congress, the administration, and the Corps have learned, the notion of optimal river management is viewed differently by different groups, and thus not easy to realize to the satisfaction of all. Attempts to tame the river for the benefit of one class of user have usually changed the river in ways that have negative consequences for other users. With changes in economies, affluence, and social preferences over time, the public has sought a changing mix in services, resulting in changing priorities for managing the river. The need to address shifting social and economic preferences, while also servicing traditional users, has posed great challenges to the Corps during this 175-year period and will continue to do so in the future.

The Corps' Restructured UMR-IWW Feasibility Study represents the most recent rendition of these efforts to manage UMR-IWW resources. The Corps encountered several analytical challenges in the course of a study process that took more than 15 years. Despite these problems and a number of serious criticisms—including some from this committee—the Corps took a major step forward by considering ecological restoration and commercial navigation in the same study. Nevertheless, as pointed out in this committee's second report, the ecosystem restoration plan's objectives are limited, stopping far short of correcting cumulative ecological changes that have resulted from construction and operation of the UMR-IWW navigation project. In its first two reports this committee noted the complexities of integrated river management and the challenges of encompassing all relevant water-related sectors within a single unifying framework. A lesson for future planning studies is that it is not sufficient to simply accumulate more information and consider additional water-related sectors in the analysis; improved planning will require careful understanding of the opportunities for trade-offs among major classes of river users and values.

The Corps' feasibility study had to address high levels of uncertainty in many of its subject areas, including waterway traffic forecasts, river responses to operational changes, and future navigation and shipping technologies and practices. These uncertainties are characteristic of all studies of this kind and were particularly prominent in the UMR-IWW Feasibility Study. The existence and nature of trade-offs among river management purposes and goals are similarly uncertain. In the interest of reducing uncertainties the committee has stressed the need for the best professional planning and analysis. But improved planning and analysis can only reduce uncertainties, not eliminate them. The possibility of making costly, inappropriate decisions based on uncertain data still exists. In a project and system like the UMR-IWW that must be operated over a long period of time, however, learning from experience—that is, applying adaptive management principles—can lead to better choices over the life of the project. To its credit the Corps has proposed a comprehensive application of adaptive management in the implementation phase, and has structured and scheduled its project proposals in a way that facilitates adaptive management. As the Corps proceeds with future UMR-IWW management decisions, an adaptive approach in which engineers, natural and social scientists, and other professionals collaborate closely with the Corps, in a two-way exchange of information and knowledge, should prove useful in improving overall knowledge of the system and in ensuring better operational decisions.

Despite the positive prospects of adaptive management for future UMR-IWW management, after some five years of interaction between the Corps and two different NRC committees, notable deficiencies within the planning study identified by these committees were never fully resolved. Both Phase I and Phase II committees concluded that the benefits of lock extensions could not be adequately evaluated without first applying nonstructural strategies for managing waterway congestion. These committees also found that the economic models used in the feasibility study to estimate the benefits of navigation improvements did not produce credible results. It would have been preferable for both of these key analytic issues to have been resolved earlier in the planning process. A firm commitment to adaptive management on the UMR-IWW leaves open the possibility that deficiencies in the planning study can be corrected prior to major investment. But adaptive management cannot be relied on to fully compensate for fundamental weaknesses in project plans. Adaptation works best when it is used to make incremental improvements in conceptually sound plans.

The UMR-IWW Feasibility Study is not the first time that the Corps has grappled with the complexity of large, multiple purpose projects affecting large populations and major ecosystems. The South Florida Comprehensive Everglades Restoration Plan, for example, is a similar situation in some respects. The process of revising the Corps' Master Manual for the Missouri River Dam and Reservoir System also bears similarities to the UMR-IWW management process. The UMR-IWW study, however, occupies a unique place in the history of such planning efforts, in that it both predates and postdates the movement of the Corps into projects with both National Economic Development and National Ecosystem Restoration purposes. The consequence of this paradigm shift was its major restructuring in 2001-2002, during which the study broadened from a conventional, single-purpose navigation improvement study to include an ecosystem restoration component. The resulting study incorporated some cutting-edge features (e.g., provisions for adaptive management, effective use of expert review, substantial stakeholder involvement) but also inherited certain outmoded or discredited elements from past planning practices (e.g., navigation benefit models, inadequate attention to nonstructural alternatives). The Corps' UMR-IWW feasibility study represents a major milestone in a long process of trying to enhance the economy of the Upper Mississippi River region, in the broad sense of conserving and rehabilitating its environmental and social features along with its engineered structures. The Corps has taken impressive strides in crossing this milestone, but as the Corps moves forward with UMR-IWW system management, a key challenge will be to retain the better features of the present plan, while correcting and strengthening its weaker elements within the context of the proposed implementation schedule.

References

Denver Post. 2005. Water decision disappointing. Denver Post Editorial, May 12, 2005.

Freeman, A.M., III. 1993. The Measurement of Environmental and Resource Values; Theory and Methods. Washington, D.C.: Resources for the Future.

Holling, C.S. (ed.). 1978. Adaptive Environmental Assessment and Management. New York, NY: John Wiley and Sons.

Jacobs, J. 2002. Broadening U.S. Water Resources Project Planning and Evaluation. Natural Resources Journal 42(1):21-31.

Kelley, R. 1998. Battling the Inland Sea: Floods, Public Policy, and the Sacramento Valley. Berkeley, Calf.: University of California Press.

Lee, K. 1999. Appraising Adaptive Management. Conservation Ecology 3(2):3. Available online at http://www.consecol.org/vol3/iss2/art3. Accessed June 6, 2005.

NRC (National Research Council). 2001. Inland Navigation System Planning: The Upper Mississippi River-Illinois Waterway. Washington, D.C.: National Academy Press.

NRC. 2002. Review Procedures for Water Resources Project Planning. Washington, DC: National Academy Press.

NRC. 2004a. Review of the U.S. Army Corps of Engineers Restructured Upper Mississippi River-Illinois Waterway Feasibility Study. Washington, D.C.: National Academy Press.

NRC. 2004b. Review of the U.S. Army Corps of Engineers Restructured Upper Mississippi River-Illinois Waterway Feasibility Study: Second Report. Washington, D.C.: National Academy Press.

NRC. 2004c. Adaptive Management for Water Resources Project Planning. Washington, D.C.: National Academy Press.

NRC. 2004d. Analytical Methods and Approaches for Water Resources Project Planning. Washington, D.C.: National Academy Press.

NRC. 2004e. River Basins and Coastal Systems Planning within the U.S. Army Corps of Engineers. Washington, D.C.: National Academy Press.

NRC. 2004f. U.S. Army Corps of Engineers Water Resources Planning: A New Opportunity for Service. Washington, D.C.: National Academy Press.

NRC. 2005. Valuing Ecosystem Services: Toward Better Environmental Decision-Making. Washington, D.C.: National Academy Press.

NWF (National Wildlife Federation) and Taxpayers for Common Sense. 2004. Crossroads: Congress, the Corps of Engineers, and the Future of America's Water Resources. Washington, D.C.: National Wildlife Federation.

Stakhiv, R., R. Cole, P. Scodari, and L. Martin. 2003. Improving Environmental Benefits Analysis in Ecosystem Restoration Planning. IWR Report 03-PS-3. Alexandria, Va.: Institute for Water Resources, U.S. Army Corps of Engineers.

Twain, M. 1883. Life on the Mississippi. Boston: James R. Osgood and Company.

USACE (U.S. Army Corps of Engineers). 2000. Planning Guidance Notebook. ER 1105-2-100, April 22, 2000. Available online at http://www.usace.army.mil/inet/usace-docs/eng-regs/er1105-1-100/tochtm. Accessed June 6, 2005.

USACE, 2005. Peer Review of Decision Documents. EC 1105-2-408. Available online at http://www.usace.army.mil/publications/eng-circulars/ec1105-2-408/toc.htm. Accessed August 4, 2005.

WRC (U.S. Water Resources Council). 1973. Water and Related Land Resources: Establishment of Principles and Standards. Washington, D.C.: U.S. Government Printing Office.

WRC (U.S. Water Resources Council). 1983. Economic and Environmental Principles and Guidelines for Water and Related Land Implementation Studies. Washington, D.C.: U.S. Government Printing Office.

White, G.F. 1971. Unpublished paper, delivered at Columbia University, March 21.

Appendix A

Water Resources Development Act 2000 Public Law No. 106-541, of the 106th Congress

SEC. 216. NATIONAL ACADEMY OF SCIENCES STUDY

(a) DEFINITIONS—In this section, the following definitions apply:

(1) ACADEMY—The term "Academy" means the National Academy of Sciences.

(2) METHOD—The term "method" means a method, model, assumption, or other pertinent planning tool used in conducting an economic or environmental analysis of a water resources project, including the formulation of a feasibility report.

(3) FEASIBILITY REPORT—The term "feasibility report" means each feasibility report, and each associated environmental impact statement and mitigation plan, prepared by the Corps of Engineers for a water resources project.

(4) WATER RESOURCES PROJECT—The term "water resources project" means a project for navigation, a project for flood control, a project for hurricane and storm damage reduction, a project for emergency streambank and shore protection, a project for ecosystem restoration and protection, and a water resources project of any other type carried out by the Corps of Engineers.

(b) INDEPENDENT PEER REVIEW OF PROJECTS—

(1) IN GENERAL—Not later than 90 days after the date of enactment of this Act, the Secretary shall contract with the Academy to study, and make recommendations relating to, the independent peer review of feasibility reports.

(2) STUDY ELEMENTS—In carrying out a contract under paragraph (1), the Academy shall study the practicality and efficacy of the independent peer review of the feasibility reports, including—

(A) the cost, time requirements, and other considerations relating to the implementation of independent peer review; and

(B) objective criteria that may be used to determine the most effective application of independent peer review to feasibility reports for each type of water resources project.

(3) ACADEMY REPORT—Not later than 1 year after the date of a contract under paragraph (1), the Academy shall submit to the Secretary, the Committee on Transportation and Infrastructure of the House of Representatives, and the Committee on Environment and Public Works of the Senate a report that includes—

(A) the results of the study conducted under paragraphs (1) and (2); and

(B) in light of the results of the study, specific recommendations, if any, on a program for implementing independent peer review of feasibility reports.

(4) AUTHORIZATION OF APPROPRIATIONS—There is authorized to be appropriated to carry out this subsection $1,000,000, to remain available until expended.

(c) INDEPENDENT PEER REVIEW OF METHODS FOR PROJECT ANALYSIS—

(1) IN GENERAL—Not later than 90 days after the date of enactment of this Act, the Secretary shall contract with the Academy to conduct a study that includes—

(A) a review of state-of-the-art methods;

(B) a review of the methods currently used by the Secretary;

(C) a review of a sample of instances in which the Secretary has applied the methods identified under subparagraph (B) in the analysis of each type of water resources project; and

(D) a comparative evaluation of the basis and validity of state-of-the-art methods identified under subparagraph (A) and the methods identified under subparagraphs (B) and (C).

(2) ACADEMY REPORT—Not later than 1 year after the date of a contract under paragraph (1), the Academy shall transmit to the Secretary, the Committee on Transportation and Infrastructure of the House of Representatives, and the Committee on Environment and Public Works of the Senate a report that includes—

(A) the results of the study conducted under paragraph (1); and

(B) in light of the results of the study, specific recommendations for modifying any of the methods currently used by the Secretary for conducting economic and environmental analyses of water resources projects.

(3) AUTHORIZATION OF APPROPRIATIONS—There is authorized to be appropriated to carry out this subsection $2,000,000. Such sums shall remain available until expended.

Appendix B

National Research Council Board Membership and Staff

WATER SCIENCE AND TECHNOLOGY BOARD

R. RHODES TRUSSELL, *Chair,* Trussell Technologies, Inc., Pasadena, California
MARY JO BAEDECKER, U.S. Geological Survey (Retired), Vienna, Virginia
JOAN G. EHRENFELD, Rutgers University, New Brunswick, New Jersey
DARA ENTEKHABI, Massachusetts Institute of Technology, Cambridge
GERALD E. GALLOWAY, University of Maryland, College Park
PETER GLEICK, Pacific Institute for Studies in Development, Environment, and Security, Oakland, California
CHARLES N. HAAS, Drexel University, Philadelphia, Pennsylvania
KAI N. LEE, Williams College, Williamstown, Massachusetts
JAMES K. MITCHELL, Virginia Polytechnic Institute and State University (Emeritus), Blacksburg
CHRISTINE L. MOE, Emory University, Atlanta, Georgia
ROBERT PERCIASEPE, National Audubon Society, New York, New York
LEONARD SHABMAN, Resources for the Future, Washington, DC
KARL K. TUREKIAN, Yale University, New Haven, Connecticut
HAME M. WATT, Independent Consultant, Washington, DC
CLAIRE WELTY, University of Maryland, Baltimore County
JAMES L. WESCOAT, JR., University of Illinois at Urbana-Champaign
GARRET P. WESTERHOFF, Malcolm Pirnie, Inc., Fair Lawn, New Jersey

Staff

STEPHEN D. PARKER, Director
LAURA J. EHLERS, Senior Staff Officer
JEFFREY W. JACOBS, Senior Staff Officer

STEPHANIE E. JOHNSON, Senior Staff Officer
WILLIAM S. LOGAN, Senior Staff Officer
LAUREN E. ALEXANDER, Staff Officer
M. JEANNE AQUILINO, Financial and Administrative Associate
ELLEN A. DE GUZMAN, Research Associate
DOROTHY K. WEIR, Research Associate
ANITA A. HALL, Program Associate

TRANSPORTATION RESEARCH BOARD EXECUTIVE COMMITTEE

Appendix C

Committee Members and Staff Biographies

COMMITTEE MEMBERS

JOHN J. BOLAND (*chair*) is a professor emeritus in the Department of Geography and Environmental Engineering at Johns Hopkins University. His fields of research include water and energy resources, environmental economics, and public utility management. Dr. Boland has studied resource problems in more than 20 countries, has published more than 200 papers and reports, and has coauthored two books on water demand management and three others on environmental management issues. Dr. Boland is a registered professional engineer. He has served on several National Research Council (NRC) committees and boards, including the Water Science and Technology Board, of which he was a founding member (1982) and past chair (1985-1988). He is a life member of the American Water Works Association and past chairman of its Economic Research Committee. Dr. Boland received his Ph.D. degree in environmental economics from Johns Hopkins University.

PATRICK BREZONIK is a professor in the Department of Civil Engineering and the past director of the Water Resources Center at the University of Minnesota. Prior to his appointment at the University of Minnesota, Dr. Brezonik was a professor of water chemistry and environmental science at the University of Florida. His fields of research include biogeochemical processes in aquatic systems, with an emphasis on the impacts of human activity on water quality and element cycles in lakes and watersheds. He is a past member of the NRC's Water Science and Technology Board and of several NRC committees, including chair of the Committee to Revitalize Education in the Field of Limnology. He received his B.S. degree in chemistry from Marquette University and his M.S. and Ph.D. degrees in water chemistry from the University of Wisconsin-Madison.

ROBERT K. DAVIS has most recently been associated with the Institute of Behavioral Science at the University of Colorado. He is the former head of the Economic Staff in the Office of the Secretary of the U.S. Department of the Interior. His fields of research include natural resource economics, environmental policy analysis, water resources planning, and methods of benefit-cost analysis. His Ph.D. thesis is widely considered the first publication on contingent valuation, a method in wide use today to quantify environmental benefits and damages. Dr. Davis has served as an adviser to foreign governments, has served in faculty positions at several universities, and has served on the staff of Resources for the Future. Dr. Davis received his B.S. and his M.S. degrees from Ohio State University and his M.P.A. and Ph.D. degrees from Harvard University.

LEO M. EISEL is a principal engineer at Brown and Caldwell in Denver, Colorado. Dr. Eisel has more than 29 years of experience with water rights and water resources. He is the former director of the Illinois Environmental Protection Agency, the Illinois Division of Water Resources, and the U.S. Water Resources Council. He is also a past president of McLaughlin Water Engineers in Denver. Dr. Eisel has served on several National Research Council committees and has served as a member of the NRC's Water Science and Technology Board. He received his Ph.D. degree in engineering from Harvard University.

STEPHEN W. FULLER is a professor in the Department of Agricultural Economics at Texas A&M University. Dr. Fuller's fields of research focus on transportation, marketing, and international trade issues, with an emphasis on the economics of Mississippi River waterway transportation. Dr. Fuller served on the NRC Committee on Freight Transportation Needs for the 21st Century. He is author of 280 refereed journal articles and reports that focus on agricultural transportation and marketing issues. Dr. Fuller has been honored five times by the Transportation Research Forum for his research by receiving the Outstanding Paper in Rural Transportation Award. Dr. Fuller received his B.S. and M.S. degrees in agricultural economics and his Ph.D. degree in economics, all from Kansas State University.

GERALD E. GALLOWAY is a research professor and professor of engineering at the Glen L. Martin Institute, University of Maryland, College Park. Before joining the University of Maryland, he was vice president of the Enterprise Engineering Group at the Titan Corporation in Arlington, Virginia. Dr. Galloway is a former secretary of the U.S. Section of the International Joint Commission. Dr. Galloway has served as a consultant on water resources engineering and management issues to the Executive Office of the President, the World Bank, the Organization of American States, the Tennessee Valley Authority, and the U.S. Army Corps of Engineers. Dr. Galloway is a former dean of the Academic Board (chief academic officer) of the U.S. Military

Academy. Dr. Galloway holds M.S. degrees from Princeton, Penn State, and the U.S. Army Command and General Staff College. Dr. Galloway received his Ph.D. degree in geography from the University of North Carolina.

LESTER B. LAVE (IOM) is the Harry B. and James H. Higgins Professor of Economics and University Professor at Carnegie Mellon University. His fields of research include applied economics and public policy, safety goals for dams and other structures, and quantitative risk assessment. Dr. Lave chaired the NRC Committee to Review the Upper Mississippi River-Illinois Waterway Navigation System Feasibility Study. He is a member of the U.S. Environmental Protection Agency Science Advisory Board and the former president of the Society for Risk Analysis. Dr. Lave received his Ph.D. degree in economics from Harvard University.

KARIN E. LIMBURG is an associate professor at the College of Environmental Science and Forestry at Syracuse University. Her fields of research focus on the Hudson River estuary in eastern New York State. Dr. Limburg teaches a course in fisheries biology and is a co-convener of a seminar series in interdisciplinary courses in watershed ecology. She received her A.B. degree from Vassar College in ecology-conservation and biology, her M.S. degree from the University of Florida in systems ecology, and her Ph.D. degree from Cornell University in ecology and evolutionary biology.

ELIZABETH A. RIEKE is the Lohontan Basin area manager for the U.S. Bureau of Reclamation in Carson City, Nevada. Ms. Rieke is a former director of the Natural Resource Law Center, University of Colorado School of Law, and a former assistant secretary for water and science in the U.S. Department of the Interior. She has served as an associate (1987-1989) and as a partner (1989-1991) with the law firm Jennings, Strouss & Salmon. Ms. Rieke received her B.A. degree from Oberlin College and her J.D. degree from the University of Arizona.

SOROOSH SOROOSHIAN (NAE) is a distinguished professor and the director of the Center for Hydrometeorology and Remote Sensing in the Department of Civil and Environmental Engineering at the University of California, Irvine. His fields of research include surface hydrology (with an emphasis on precipitation runoff modeling), the hydrology of arid and semiarid regions, and related water resources management issues. He has served on several NRC committees, including a six-year term as the chair of the NRC Global Energy and Water Cycle Experiment (GEWEX) Panel. Dr. Sorooshian was elected to the National Academy of Engineering in 2003. Dr. Sorooshian received his B.S. degree from California State Polytechnic University and his M.S. and Ph.D. degrees from the University of California, Los Angeles.

RICHARD E. SPARKS is Director of Research, National Great Rivers Research and Education Center, Alton, Illinois, which is a partnership of the University of Illinois at Urbana-Champaign, Lewis and Clark Community College, the Illinois Natural History Survey, and other institutions concerned with management, education, and research on rivers and watersheds. He currently researches options for managing invasive aquatic species and restoring or naturalizing large floodplain rivers. He continues to be affiliated with the University of Illinois, where he formerly directed the Illinois Water Resources Center, and with the Illinois Natural History Survey, where he directed the Large River Research Program on the Upper Mississippi River system. He has served on several NRC committees, including the Committee on Aquatic Restoration and the Committee to Assess U.S. Army Corps of Engineers Water Resources Project Planning Procedures. In Argentina, Brazil and India he provided advice on management of floodplain ecosystems and large rivers. He received a B.A. degree from Amherst College, his M.S. degree in biology from the University of Kansas, and his Ph.D. degree in biology from the Virginia Polytechnic Institute and State University.

STAFF

JEFFREY W. JACOBS is a senior program officer at the National Research Council's Water Science and Technology Board. Dr. Jacobs' research interests include policy and organizational arrangements for water resources management and the use of scientific information in water resources decision making. He has studied these issues extensively both in the United States and in mainland Southeast Asia. Since joining the NRC in 1997, he has served as the study director of 14 NRC committees. He received his B.S. degree from Texas A&M University, his M.A. degree from the University of California, Riverside, and his Ph.D. degree from the University of Colorado.

JOSEPH R. MORRIS is a senior program officer with the National Academies' Transportation Research Board (TRB). On the staff of TRB's Studies and Information Services Division since 1983, Mr. Morris has participated in studies of freight transportation, highway safety, transportation finance, highway design standards, and transportation and air quality. He received his B.A. from Oberlin College, his master of city and regional planning degree from Harvard University, and his M.S. degree from the University of Chicago.